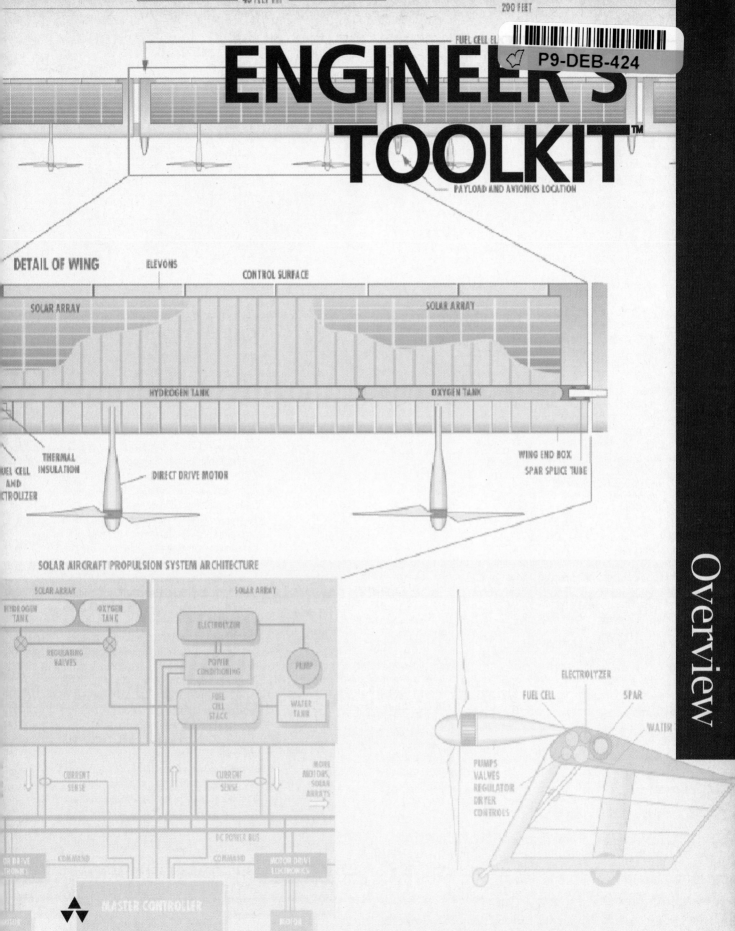

40 FEET TYP

200 FEET

P9-DEB-424

FUEL CELL EL

PAYLOAD AND AVIONICS LOCATION

ENGINEER'S TOOLKIT™

Overview

DETAIL OF WING

ELEVONS

CONTROL SURFACE

SOLAR ARRAY

SOLAR ARRAY

HYDROGEN TANK

OXYGEN TANK

THERMAL INSULATION

DIRECT DRIVE MOTOR

WING END BOX

SPAR SPLICE TUBE

FUEL CELL AND CTROLIZER

SOLAR AIRCRAFT PROPULSION SYSTEM ARCHITECTURE

SOLAR ARRAY

SOLAR ARRAY

HYDROGEN TANK

OXYGEN TANK

ELECTROLIZER

REGULATING VALVES

POWER CONDITIONING

PUMP

FUEL CELL STACK

WATER TANK

CURRENT SENSE

CURRENT SENSE

MORE MOTORS, SOLAR ARRAYS

DC POWER BUS

COMMAND

COMMAND

MOTOR DRIVE ELECTRONICS

MASTER CONTROLLER

ELECTROLYZER

FUEL CELL

SPAR

WATER

PUMPS
VALVES
REGULATOR
DRYER
CONTROLS

Addison-Wesley Publishing Company, Inc.

Menlo Park, California · Reading, Massachusetts · New York · Don Mills, Ontario
Wokingham, U.K. · Amsterdam · Bonn · Paris · Milan · Madrid · Sydney
Singapore · Tokyo · Seoul · Taipei · Mexico City · San Juan, Puerto Rico

Executive Editor: Dan Joraanstad
Acquisitions Editor: Denise Penrose
Marketing Manager: Mary Tudor
Developmental Editors: Deborah Craig,
Jeannine Drew, Kate Hoffman,
Shelly Langman
Assistant Editor: Nate McFadden
Senior Production Editor: Teri Holden
Production Editors: Jean Lake,
Gail Carrigan, Catherine Lewis
Supplements Production Editor: Teresa
Thomas
Photo Editor: Lisa Lougee
Copy Editors: Barbara Conway, Robert
Fiske
Proofreader: Holly McLean-Aldis
Marketing Coordinator: Anne Boyd
Cover Design: Yvo Riezebos
Text Design: Side by Side Studios
Technology Support: Craig Johnson
Composition: Side by Side Studios, Fog
Press, London Road Design, Progressive,
The Printed Page
Manufacturing Coordinator: Janet Weaver
Printing and Binding: R. R. Donnelley

Cover and Overview Photo Credits
Cover: Photo ©James Caccavo/Zuma
Images; background illustration
©Ian Worpole
Photo 1: Courtesy of NASA
Photo 2: Courtesy of Jet Propulsion
Laboratory
Photo 3: ©David Parker/SPL/Photo
Researchers, Inc.
Photo 4: ©Brownie Harris/The Stock
Market
Photo 5: Courtesy of M.E. Raichle, Wash.
Univ., St. Louis
Photo 6: ©Chuck O'Rear/Westlight
Photo 7: Courtesy of Lockheed. Photo by
Russ Underwood.
Photo 8: ©Roger Ressmeyer/Starlight
Photo 9: ©George Haling/Photo
Researchers, Inc.
Photo 10: Courtesy of Keith Wood,
Promega, Inc.

The programs, worksheets, and examples
presented in this book have been
included for their instructional value.
They have been tested with care but are
not guaranteed for any particular purpose. The publisher does not offer any
warranties or representations, nor does it
accept any liabilities with respect to the
programs, worksheets, or examples.

ISBN: 0-8053-6335-1

Addison-Wesley Publishing Company, Inc.
2725 Sand Hill Road
Menlo Park, CA 94025

COVER STORY

Pictured on the cover of The Engineer's
Toolkit is the Pathfinder—a "solar-powered
flying wing" designed for low-speed, high-
altitude flight. With a wing span compara-
ble to a Boeing 737, it weighs in at just 400
pounds and has no rudders, no fins, no tail,
and no pilot! The Pathfinder is one of a
series of solar planes developed and built
by Dr. Paul MacCready and his team of
engineers at AeroVironment Inc., in Simi
Valley, California. Engineers at the
Lawrence Livermore National Laboratory in
Livermore, California, designed, engineered, and continue to administer the
Pathfinder solar plane. This laboratory also
is designing the next iteration of solar
planes, the Helios (plans for which appear
behind the photograph of the Pathfinder).
With the Helios, engineers hope to come
even closer to realizing the dream of "eter-
nal flight"; it will include an on-board
energy storage system that can provide the
energy needed during night flight.

As with most contemporary engineering projects, designing solar planes requires
the efforts of engineers from a variety of
disciplines—aeronautical, computer, electrical, environmental, and mechanical, to
name a few. Still other teams of engineers
are needed to design on-board equipment
to support specific missions, such as monitoring dangerous weather systems or tracking the release of toxins into the atmosphere.

Contemporary design examples such as
these are presented throughout The Engineer's Toolkit, highlighting the interdisciplinary teamwork that characterizes engineering today.

Tools for a New Curriculum

The Engineer's Toolkit is not a conventional textbook. It was inspired by the needs of instructors like you, who are engaged in developing a new curriculum in introductory engineering courses. They are searching for new ways to prepare, motivate, and engage first-year students. They want to create a link for their students between the prerequisite math and science courses and the wide range of engineering disciplines that build on that knowledge. These instructors also want to ensure that their students master the skills of team-building, communications, and computer use—skills they need to solve problems successfully in subsequent courses and in the real world of work. You and your colleagues are also experimenting with hands-on design projects so students understand that design is a process and that, fundamentally, engineering means solving problems.

Universities and colleges are responding in unique ways to the changing landscape of introductory engineering. This very uniqueness creates a new challenge when you are searching for the right text to support your unique course. The Engineer's Toolkit takes on that challenge. You choose from a rich set of course materials that introduce fundamental concepts of engineering and teach essential skills and tools. Each tool is presented as a single module. You determine which modules will best satisfy your course goals, and Addison-Wesley binds those modules into the exact book your students need.

Especially written and designed for The Engineer's Toolkit, the modules present a consistent teaching methodology adapted from the work of Delores Etter, author of the spreadsheet and Fortran Toolkit modules, as well as Structured Fortran 77 for Scientists and Engineers. Each of the Toolkit authors has applied

Dr. Etter's five-step problem-solving process to a wide variety of programming languages and application programs. A consistent approach, style, level, and tone means you and your students don't have to switch gears every time you begin to teach or learn a new tool.

Here are the six key pedagogical features you'll find in the Toolkit modules that teach programming languages or software tools:

- **The five-step problem-solving process** is explained and illustrated in terms of the particular language or software tool being taught. It is then used throughout the module in applications, numbered examples, and end-of-chapter exercises or problems.
- **Applications** based on the Ten Great Engineering Achievements and representing a wide variety of engineering disciplines demonstrate the five-step problem-solving process.
- **"What If?" problems** immediately follow applications in the software tools modules, asking students to modify assumptions, data, or variables in the application and to solve the new problems that result.
- **Numbered examples** demonstrate key elements of a language or application program by providing fully worked-out solutions.
- **"Try It!" exercises** test students' knowledge of sections within a chapter and frequently require work at the computer.
- **End-of-chapter material** includes summaries of essential points, a key word list, and a set of exercises or problems that gradually increase in complexity.
  These pedagogical features are also described from a student's point of view in the section "The Toolkit Methodology."

How To Design Your Custom Textbook

A sample course goal: To introduce engineering, teach a programming language, word processing and CAD techniques.

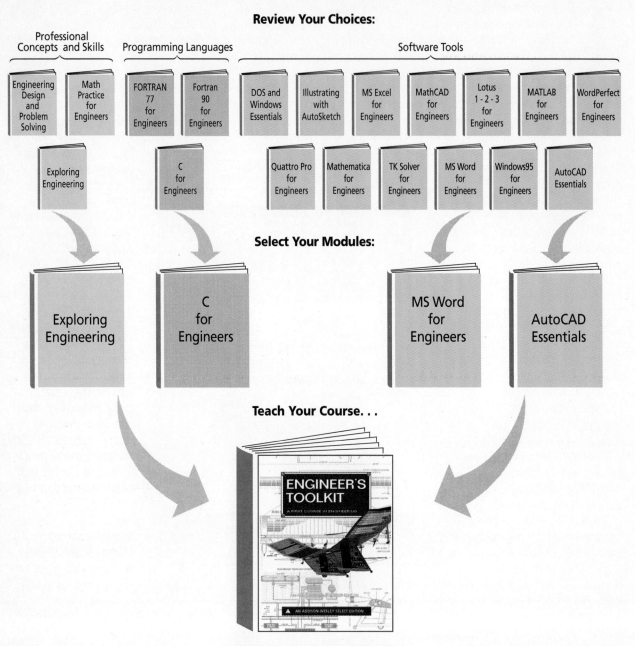

Review Your Choices:

Professional Concepts and Skills

- Engineering Design and Problem Solving
- Math Practice for Engineers
- Exploring Engineering

Programming Languages

- FORTRAN 77 for Engineers
- Fortran 90 for Engineers
- C for Engineers

Software Tools

- DOS and Windows Essentials
- Illustrating with AutoSketch
- MS Excel for Engineers
- MathCAD for Engineers
- Lotus 1-2-3 for Engineers
- MATLAB for Engineers
- WordPerfect for Engineers
- Quattro Pro for Engineers
- Mathematica for Engineers
- TK Solver for Engineers
- MS Word for Engineers
- Windows95 for Engineers
- AutoCAD Essentials

Select Your Modules:

- Exploring Engineering
- C for Engineers
- MS Word for Engineers
- AutoCAD Essentials

Teach Your Course. . .

ENGINEER'S TOOLKIT

A FIRST COURSE IN ENGINEERING

AN ADDISON-WESLEY SELECT EDITION

With Your Custom Textbook!

DESIGN YOUR OWN TOOLKIT

The *Toolkit* menu consists of three modules that teach programming languages, thirteen modules that teach software tools, and three modules that focus on core engineering concepts and skills—*Engineering Design and Problem Solving, Exploring Engineering,* and *Math Practice for Engineers.* You can mix and match as your course demands.

In a course that combines an overview of engineering disciplines with essential computer applications and the basic design process, you might, for example, combine *Exploring Engineering, Engineering Design and Problem Solving, Quattro Pro for Engineers,* and *Microsoft Word for Engineers* for a book of about 460 pages.

If you plan to teach programming and introduce your students to design and basic CAD techniques, you might combine *Engineering Design and Problem Solving, C for Engineers,* and *AutoCAD Essentials* for a book of about 550 pages.

The *Toolkit* modules are also available separately, and each custom text includes this Overview, which introduces instructors and students to the *Toolkit* methodology.

FUTURE MODULES

Each year Addison-Wesley will release additional modules to keep pace with the development in introductory engineering courses. We welcome your suggestions for future modules in *The Engineer's Toolkit*. Please correspond with us at the Internet address

toolkit@aw.com

or by regular mail at

Toolkit
Addison-Wesley Publishing
Company, Inc.
2725 Sand Hill Road
Menlo Park, CA 94025

Toolkit information is also available on the World Wide Web at

http://www.aw.com/cseng/toolkit/

ELECTRONIC SUPPLEMENTS AND SUPPORT MATERIAL

Addison-Wesley offers a range of support materials for *The Engineer's Toolkit*, both printed and electronic. We welcome your comments on the effectiveness of these materials and your suggestions for additional supplements.

Supplements via the Internet A selection of supplementary material for *The Engineer's Toolkit*, including transparency masters, is available on the Internet via anonymous FTP. The URL for accessing this material is ftp://aw.com/cseng/toolkit.

Instructor's Guides Instructor's Guides for the modules you choose will be provided upon adoption of *The Engineer's Toolkit*.

Each Instructor's Guide opens with an overview of the module subject matter, suggested methods of instruction, comments on tests and quizzes, and a general discussion of software version and platform issues (if applicable).

Each Instructor's Guide also describes how to order the module topics to support various syllabi.

A chapter-by-chapter section presents teaching strategies, points to emphasize and special challenges for each chapter, solutions to the end-of-chapter problems, and additional problems and solutions.

Student Data Files Student Data Files, including the data, applications, and program files that support end-of-chapter problems in specific modules, will be available in the following ways:

- ◆ on the disk that includes the Instructor's Guide
- ◆ via our FTP site: ftp://aw.com/cseng/toolkit/igs/
- ◆ via our Toolkit home page on the World Wide Web: http://www.aw.com/cseng/toolkit/

Modules with Data Files include *AutoCAD Essentials, Illustrating with AutoSketch, FORTRAN 77 for Engineers, Fortran 90 for Engineers, C for Engineers, Quattro Pro for Engineers, Lotus 1-2-3 for Engineers, Microsoft Excel for Engineers, TK Solver for Engineers,* and *Mathematica for Engineers.*

Errata for Published Modules: Errata notices for published modules will be available online at: http://www.aw.com/cseng/toolkit/

FIRST ASSIGNMENT

The next section of this Overview is directed to your students. It explains the teaching and learning strategies adopted by our authors throughout the *Toolkit*. The final section introduces students to the Ten Great Engineering Achievements. We invite you to read on and hope you will assign this section to your students early in your course.

THE ENGINEER'S TOOLKIT MODULES

	Title	Author
Professional Concepts and Skills	Engineering Design and Problem Solving	Steve Howell
	Exploring Engineering	Joe King
	Math Practice for Engineers	Joe King
Software Tools	AutoCAD Essentials	Melton Miller
	DOS and Windows Essentials	Gerald Lemay
	Illustrating with AutoSketch	Gordon Snyder
	Lotus 1-2-3 for Engineers	Delores Etter
	MathCAD for Engineers	Joe King
	Mathematica for Engineers	Henry Shapiro
	MATLAB for Engineers	Joe King
	Microsoft Excel for Engineers	Delores Etter
	Microsoft Word for Engineers	Sheryl Sorby
	Quattro Pro for Engineers	Delores Etter
	TK Solver for Engineers	Robert J. Ferguson
	Windows95 for Engineers	Gordon Snyder
	WordPerfect for Engineers	Sheryl Sorby
Programming Languages	C for Engineers	Kenneth Collier
	FORTRAN 77 for Engineers	Delores Etter
	Fortran 90 for Engineers	Delores Etter

The Toolkit Methodology

Welcome to *The Engineer's Toolkit!* This book has been especially created to support your work in what is probably your first course in Engineering. Unlike other textbooks you have studied, *The Engineer's Toolkit* was customized by your instructor to include the exact material you need, and only the material you need. In essence, *The Engineer's Toolkit* is a collection of modules that teach engineering concepts and skills, software tools, and programming languages. Your instructor has selected the appropriate modules for your course, and Addison-Wesley has bound those modules into this custom book.

Introduction. Each application is fully described and explained so that you have sufficient information to complete step 1.

A GENERAL PROCESS FOR SOLVING PROBLEMS

A key feature of *The Engineer's Toolkit* is its emphasis on developing problem-solving skills. Problem solving is one of the foundations of all engineering activity. In *The Engineer's Toolkit* you'll find a five-step method for solving the problems given in each module. Some engineers will tell you they use a nine-step process; others can condense their process down to four. There's nothing magic about the number, but you will find that learning and following a consistent method for solving problems will

make you an efficient student and a promising graduate. Each application program or programming language module builds on this general problem-solving method:

1. Define the problem.
2. Gather information.
3. Generate and evaluate potential solutions.
4. Refine and implement the solution.
5. Verify the solution through testing.

---

66   QUATTRO PRO FOR ENGINEERS

and choosing it. An icon will be on this page for each graph book. Choose the icon of the graph you want to rename with and select Properties|Current Object. The Name dialog box t in which you may type the new name of the graph and then c
   Saving and printing graphs is similar to saving and print sheets. The graph is automatically saved when you save the file. You can print the graph either by itself or with the spre print the graph by itself, select the graph using Graph|Edi choose File|Print. If you have placed the graph as a floating g spreadsheet, it is printed when you print the spreadsheet.

 Try It   Change the name of the GRAPH1 graph in the FILTER1 spreads NALIN. Then print the graph.

Application 1   **QUALITY CONTROL**

Manufacturing Engineering
In a manufacturing or assembly plant, quality control receives tion. One of the key responsibilities of a quality control engine lect accurate data on the quality of the product being manufa data can be used to identify the problem areas in the assemb the materials being used in the product.

Circuit Board Defects
In this application, information collected over a one-year peric specify both the type of defects and the number of defects det assembly of printed circuit boards. These defects have been four categories: board errors, chip errors, processing errors, tion errors. Board errors are typically caused by defects in th of the printed circuits. Chip errors are caused by defective in cuit (IC) chips that are added to the board; these IC chips inclu chips, microprocessor chips, and digital filter chips. Processin typically caused by errors in inserting the chips in the board; is often done by manufacturing robots, and the robot progra be incorrect, or the chips being inserted can be packaged i order. Connection errors are solder errors that occur when th through the solder machine; these errors can be caused by board or an incorrect solder temperature.

Spreadsheet for a Quality Analysis Report
You want to develop a spreadsheet that summarizes the qua data that has been collected each month for a year. This data number of defects in each of the four categories of defects dis summary report should compute totals and percentages fo year and defect totals for each quarter. Later in this chapter the data in the spreadsheet to generate pie, bar, and line grap

---

184   C FOR ENGINEERS

```
void init_array(struct sample_type array[MAX_RAINFALLS][MAX_SITES])
{
    int row, col;          /* Loop control variables */

    for (row=0; row < MAX_RAINFALLS; ++row)
        for (col=0; col < MAX_SITES; ++col){
            array[row][col].date.day = 0;
            array[row][col].date.month = 0;
            array[row][col].date.year = 0;
            array[row][col].time.hour = 0;
            array[row][col].time.minutes = 0;
            array[row][col].h_concentration = 0.0;
            array[row][col].ph_level = 0.0;
        }
}
```

Carefully examine the definition of init_array(). Notice that the parameter array is declared as a two-dimensional array of structures. In the body of this function access is made to elements within array using the familiar subscript notation. Once a particular element has been accessed in this manner, the fields of that structure are accessed using the dot operator. Whenever the accessed field is itself a structure, its fields are accessed using a second dot operator, as in the statement

```
array[row][col].date.month = 0;
```

Because the dot is an operator, it can be combined with other operators in this way.
   There are many powerful ways to use structures in a program, and this section provides you with only an introduction. In the following application you will learn how structures can be used to extend the mathematical power of *C* to include complex arithmetic.

Application 1   **FREQUENCY DISTRIBUTION GRAPHING**

Industrial Engineering
Quality-control engineers monitor the quality of an automated production line by tracking the number of defective parts coming off the line within a particular period. If the frequency of defective parts rises dramatically for a given period, the engineer is alerted that a problem exists and can take action to fix the problem. Such frequencies can be depicted using a bar graph such as the one shown in Figure 7-5. In this graph, the horizontal axis represents the number of defects detected and the vertical axis represents data collection periods.

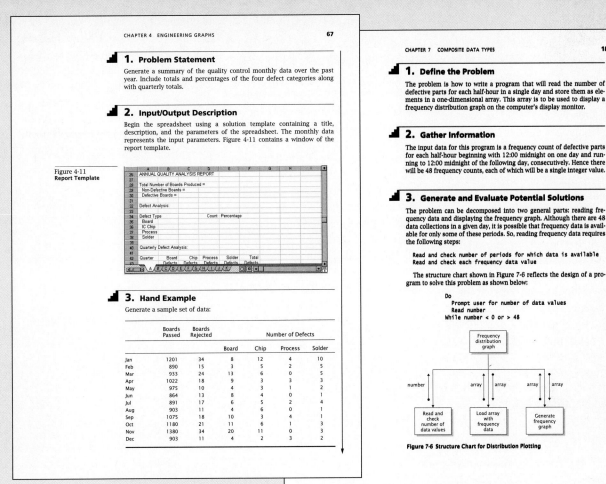

1. Problem Statement

Generate a summary of the quality control monthly data over the past year. Include totals and percentages of the four defect categories along with quarterly totals.

2. Input/Output Description

Begin the spreadsheet using a solution template containing a title, description, and the parameters of the spreadsheet. The monthly data represents the input parameters. Figure 4-11 contains a window of the report template.

Figure 4-11
Report Template

3. Hand Example

Generate a sample set of data:

	Boards Passed	Boards Rejected	Number of Defects			
			Board	Chip	Process	Solder
Jan	1201	34	8	12	4	10
Feb	890	15	3	5	2	5
Mar	933	24	13	6	0	5
Apr	1022	18	9	3	3	3
May	975	10	4	3	1	2
Jun	864	13	8	4	0	1
Jul	891	17	6	5	2	4
Aug	903	11	4	6	0	1
Sep	1075	18	10	3	4	1
Oct	1180	21	11	6	1	3
Nov	1380	34	20	11	0	3
Dec	903	11	4	2	3	2

From *Quattro Pro® for Engineers*

1. Define the Problem

The problem is how to write a program that will read the number of defective parts for each half-hour in a single day and store them as elements in a one-dimensional array. This array is to be used to display a frequency distribution graph on the computer's display monitor.

2. Gather Information

The input data for this program is a frequency count of defective parts for each half-hour beginning with 12:00 midnight on one day and running to 12:00 midnight of the following day, consecutively. Hence there will be 48 frequency counts, each of which will be a single integer value.

3. Generate and Evaluate Potential Solutions

The problem can be decomposed into two general parts: reading frequency data and displaying the frequency graph. Although there are 48 data collections in a given day, it is possible that frequency data is available for only some of these periods. So, reading frequency data requires the following steps:

Read and check number of periods for which data is available
Read and check each frequency data value

The structure chart shown in Figure 7-6 reflects the design of a program to solve this problem as shown below:

```
Do
    Prompt user for number of data values
    Read number
While number < 0 or > 48
```

Figure 7-6 Structure Chart for Distribution Plotting

From *C for Engineers*

Step 1. In both applications you are asked to **define the problem.** The introductory description gives lots of clues to help you.

Step 2. This step asks you to **gather information** that you need to propose a solution. In these applications you need to prepare the data that will be used to generate quality analysis reports.

Step 3. You are now ready to **generate and evaluate potential solutions.** In C you create a structure chart to illustrate the design of a program that can solve the problem. In Quattro Pro, the hand example shows how the gathered data will be used to determine the algorithm in step 4.

A SPECIFIC FIVE-STEP PROCESS

Each module adapts this general method and refines it according to the kinds of problems solved by the tool or language being taught. Chapter 1 of each programming language and application program module describes the specific five-step process used in that module.

As you work through *The Engineer's Toolkit*, you'll find that this consistent approach makes it easier to solve new problems. For instance, step 1 of the five-step problem-solving process calls for the same kind of thinking process whether you are using a programming language like Fortran 90, a computer-aided design package like AutoCAD, or an equation solver like MATLAB.

We illustrate the five-step problem-solving process with a pair of applications from *The Engineer's Toolkit*. Both of the examples presented here deal with the collection and tracking of data related to quality control. Each has been fully worked out using the five-step process. Follow these steps to see how easy it is to learn this problem-solving process.

APPLICATIONS

These two applications are among hundreds in which *Toolkit* authors demonstrate the five-step problem-solving method. As you gain experience using this method with various software tools and languages, you'll find you can approach new problems with confidence, and you'll begin to identify the appropriate tool or language for the problem at hand. Learning to choose the right tool for a specific engineering problem is an important part of your education.

Many of the applications in *The Engineer's Toolkit* are based on the Ten Great Engineering Achievements chosen by the National Academy of Engineering to celebrate its silver anniversary in 1989. Studying these applications will help you see the kinds of problems faced by engineers from different disciplines and better understand how large problems are broken into smaller solvable problems. This Overview concludes with a description of the Ten Great Engineering Achievements.

Step 4. In this step you write a C program based on the structure chart and algorithms developed in the previous steps. You will now **refine and implement the solution.** With Quattro Pro, this step means developing the formulas that will be used to compute the values listed in the summary report.

Step 5. In the final step you **verify the solution through testing.** In C this involves entering a variety of values (value testing) to confirm that the program generates valid output. In Quattro Pro the spreadsheet is tested with several sets of data to verify the accuracy of the computations. Accuracy is confirmed by comparing the spreadsheet calculations to the values determined in the hand example.

---

4. Algorithm Development

The spreadsheet now contains everything except the formulas for computing the summary information for the report. Develop the formulas in the order needed to compute the values. Also, try to minimize the number of computations. For example, since you generate error sums by quarters, add the quarterly sums to get yearly sums instead of adding all the monthly sums to get yearly sums. It would be good practice to verify each of these formulas by referring to Figures 4-10 and 4-11.

D29	@SUM(D11..D22)	Total non-defective boards
D30	@SUM(E11..E22)	Total defective boards
E28	+D29+D30	Total boards
E29	+D29/E28	Percent non-defective boards
E30	Copy from E29	Percent defective boards
B44	@SUM(G11..G13)	Quarter I board defects
B45	@SUM(G14..G16)	Quarter II board defects

5. Testing

An important part of developing a spreadsheet is testing it with several sets of data to verify the accuracy of the computations. Using the sample set of data from the hand example, you can easily check the accuracy of the spreadsheet calculations by comparing them to the hand example.

You should make minor changes in this data and check the report to be sure that corresponding changes occur in the report summary. In this report you want to be sure that the report would be generated correctly if there were no errors in one of the categories. Be sure to change the corresponding sums of boards rejected. The corresponding report generated, shown in Figure 4-12, shows that there were no process defects during any of the four quarters.

Figure 4-12
Report with No Processing Defects

ANNUAL QUALITY ANALYSIS REPORT					
Total Number of Boards Produced =		12,423			
Non-Defective Boards =	12,217	98.34%			
Defective Boards =	206	1.66%			
Defect Analysis:					
Defect Type		Count	Percentage		
Board		100	48.54%		
IC Chip		66	32.04%		
Process		0	0.00%		
Solder		40	19.42%		
Quarterly Defect Analysis:					
Quarter	Board Defects	Chip Defects	Process Defects	Solder Defects	Total Defects
I	24	23	0	20	67
II	21	10	0	6	37
III	20	14	0	6	40
IV	35	19	0	8	62

---

4. Refine and Implement a Solution

The structure chart and algorithms developed in the previous section are implemented as the program in Example 7-10.

5. Verify and Test the Solution

To properly test this program, you should enter a variety of values for first input, including 0, 48, values below 0, values above 48, and valid values. Selecting values along the boundaries is known as boundary value testing. You should do the same for the actual frequency values. Given the input values −3 (error) 55 (error) 10−9 (error) 15 16 13 14 20 9 96 (error) 19 14 19 20, the output of this program is

```
--------------------------------------------------
 1 | **************
 2 | ****************
 3 | *************
 4 | **************
 5 | ********************
 6 | *********
 7 | ********************
 8 | **************
 9 | ********************
10 | ********************
 --- 5---10---15---20---25---30---35---40---45---50
```

This chapter introduced you to one-, two- and three-dimensional arrays, as well as structures. Arrays and structures offer a cohesive means of storing composite collections of data. Arrays are used for grouping items of the same type and meaning, whereas structures allow you to group related items with different types and meanings. You learned how to declare and initialize arrays and manipulate the elements in an array. You also learned how to define a structure, declare structure variables, and manipulate the fields in a structure. Finally, you learned that these composite data types can be combined to allow you to create arrays of structures, structures containing arrays, and structures containing other structures.

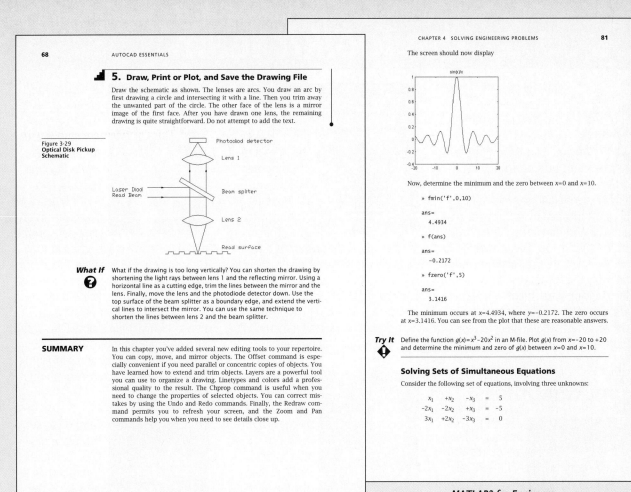

MATLAB® for Engineers

AutoCAD® Essentials

ENGINEERING DISCIPLINES

In *AutoCAD Essentials,* the author presents applications from mechanical and materials engineering, civil and structural engineering, electrical and electronics engineering, and optical engineering. *MATLAB for Engineers* includes applications from electrical and computer engineering. You can find the listing of application problems in Chapter 1 of each software tool or programming language module. The table below lists several applications from the FORTRAN 77 module.

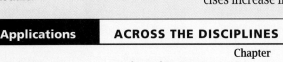

Applications	ACROSS THE DISCIPLINES	Chapter
Stride Estimation	Mechanical Engineering	2
Light Pipes	Optical Engineering	3
Sonar Signals	Acoustical Engineering	4
Wind Tunnels	Aerospace Engineering	5
Oil Well Production	Petroleum Engineering	6
Simulation Data	Electrical Engineering	7

❓ "What If" Problems

These problems immediately follow an application. Often you are asked to test the model developed in the application with different data or different assumptions. Working through these problems ensures you fully understand the application.

❗ "Try It!" Exercises

Each chapter contains several "Try It!" exercises. A "Try It!" is a short set of exercises that tests your understanding of the material. These exercises increase in complexity over the course of each module, and if you try out each one, you'll find you'll master the material faster than you expected.

NUMBERED EXAMPLES

Numbered examples appear in all the programming language and some software tool modules. These examples are designed to illustrate specific features of the language or software. Working through these examples is essential, especially those that offer two solutions and a discussion of the differences between them.

EXERCISES/PROBLEMS

Every module includes end-of-chapter exercises or problems that increase in difficulty and test your knowledge of the chapter material. Make sure you try your hand at the excercises that require you to use the five-step method to find solutions.

The *Toolkit* Team

As you consider the Ten Great Engineering Achievements explored at the end of this Overview, you'll notice that many require contributions from several different engineering disciplines. Today, significant projects can only be accomplished by teams of professionals. And that's true of *the Engineer's Toolkit,* too. The team that has helped to create the *Toolkit* includes not only the authors but also the focus group participants who honed and directed the *Toolkit* concept and the reviewers who helped develop the individual manuscripts.

TOOLKIT AUTHORS

Delores Etter, author of five modules, *FORTRAN 77 for Engineers, Fortran 90 for Engineers, Lotus 1-2-3 for Engineers, QuattroPro for Engineers, and Microsoft Excel for Engineers,* is a professor of electrical and computer engineering at the University of Colorado at Boulder. Dr. Etter has helped shape *The Engineer's Toolkit* from its initial conception, contributing the five-step problem-solving process and key pedagogical features that were successfully tested in her earlier Addison-Wesley texts, such as *Structured FORTRAN 77 for Engineers and Scientists, Lotus 123: A Software Tool for Engineers,* and *Quattro Pro: A Software Tool for Engineers.*

Ken Collier, Assistant Professor of Computer Science in the Department of Computer Science and Engineering at Northern Arizona University in Flagstaff, is the author of *C for Engineers.* Professor Collier teaches courses in C, C++, software engineering, and engineering design. His areas of research include software engineering, software design methodologies, and artificial intelligence.

R. J. Ferguson, the author of *TK Solver for Engineers,* is a professor of mechanical engineering at the Royal Military College of Canada. He teaches courses in stress analysis and computer-aided design. Other publications by Professor Ferguson include texts in the fields of fracture mechanics, noncircular gearing, vehicle transmissions, and engineering education.

Steve Howell, Associate Professor of Engineering in the mechanical engineering department at Northern Arizona University in Flagstaff, Arizona, is the author of *Engineering Design and Problem Solving.* He teaches a course titled Introduction to Engineering Design and Graphics. Professor Howell's areas of research are computer-aided design and manufacturing, thermodynamics, and heat transfer.

Joe King, the author of *MathCAD for Engineers, MATLAB for Engineers, Exploring Engineering,* and *Math Practice for Engineers,* is an associate professor of electrical engineering at the University of the Pacific in Stockton, California. He teaches courses in electrical engineering, advanced digital design, local area networks, neural networks, machine vision, C++, and microprocessor applications. He conducts research in the areas of neural networks and microprocessor applications.

Gerald Lemay, Professor of Electrical and Computer Engineering at the University of Massachusetts, Dartmouth, is the author of *DOS and Windows Essentials.* He teaches the Science of Engineering for honors students and Computer Tools for Engineers. Professor Lemay does research in renewable energy.

Melton Miller, author of *AutoCAD Essentials,* is an associate professor of civil engineering and assistant dean of the College of Engineering at the University of Massachusetts, Amherst. He teaches courses in Pascal, Fortran, Lotus, MathCAD, and AutoCAD. He also teaches courses in the the design of reinforced concrete structures.

Henry Shapiro is the author of *Mathematica for Engineers.* He is an associate professor of computer science at the University of New Mexico, where he teaches courses in computer programming and mathematical foundations of computer science. Professor Shapiro conducts research in the area of algorithm design. He is also active in curriculum development and accreditation of computer science programs.

Sheryl Sorby is the author of *WordPerfect for Engineers* and *Microsoft Word for Engineers.* She is an assistant professor of civil and environmental engineering at Michigan Technological University in Houghton, where she teaches courses in freshman engineering and computer skills. Professor Sorby conducts research in structural engineering.

Gordon Snyder, author of *Illustrating with AutoSketch* and *Windows 95 Essentials,* is an associate professor and department co-chair of the departments of electronics systems engineering, computer systems engineering, and laser electro-optics technology at Springfield Technical Community College in Springfield, Massachusetts. He teaches a course titled Introduction to Computer-Aided Engineering Technology.

REVIEWERS AND FOCUS GROUP PARTICIPANTS

Instructors throughout the country attended focus groups to help us identify key trends in engineering education. Over 100 reviewers contributed to the development of manuscripts for *The Engineer's Toolkit.* We gratefully acknowledge all their contributions.

Teresa Adams, University of Wisconsin-Madison · Anjum Ali, Mercer University · Abbas Aminmansour, Pennsylvania State University · A. Adnan Aswad, University of Michigan-Dearborn · Stormy Attaway, Boston University · James E. Bailey, Arizona State University · Betty Barr, University of Houston · O. Barron, University of Tennessee at Martin · Charles Beach, Florida Institute of Technology · Bill Beckwith, Clemson University · Charlotte Behm, Mission College · Lynn Bellamy, Arizona State University · Bhushan L. Bhatt, Oakland University · Steve Borgelt, University of Missouri · Joseph M. Bradley, United States Navy · Robert Brannock, Macon College · Gus Brar, Delaware Community College · David Bricker, Oakland University · Matthew Calame, Cosumnes River College · Christopher Carroll, University of Minnesota-Duluth · R. J. Coleman, University of North Carolina-Charlotte · James Collier, Virginia Polytechnic Institute and State University · Kenneth Collier, Northern Arizona University · Tom Cook, Mercer University · John Barrett Crittenden, Virginia Polytechnic Institute and State University · Barry Crittendon, Virginia Polytechnic Institute and State University · James Cunningham, Embry-Riddle Aeronautical University · Janek Dave, University of Cincinnati · Tim David, Leeds University · Al Day, Iowa State University · Jack Deacon, University of Kentucky · Bruce Dewey, University of Wyoming · Julie Ellis, University of Southern Maine · Beth Eschenbach, Humboldt State University · David Fletcher, University of the Pacific · Wallace Fowler, University of Texas-Austin · Jim Freeman, San Jose State University · Ben Friedman, Wolfram Research, Inc. · Paul Funk, University of Evansville · David Gallagher, Catholic University of America · Byron Garry, South Dakota State University · Larry Genalo, Iowa State University · Alan Genz, Washington State University · Johannes Gessler, Colorado State University · John J. Gilheany, Catholic University of America · Oscar R. Gonzalez, Old Dominion University · Yolanda Guran, Oregon Institute of Technology · Nicolas Haddad, New Mexico Institute of Mining and Technology · John Hakola, Hofstra University · Fred Hart, Worcester Polytechnic University · David Hata, Intel Corporation · Frank Hatfield, Michigan State University · Jim D. Helmert, Eastern Oklahoma State University · Herman W. Hill, Ohio University · Deidre A. Hirschfeld, Virginia Polytechnic Institute and State University · Margaret Hoft, University of Michigan-Dearborn · Henry Horwitz, Dutchess Community College · Peter K. Imbrie, Texas A & M University · Jeanine Ingber, University of New Mexico · Scott Iverson, University of Washington · Denise Jackson, University of Tennessee · Gerald S. Jakubowski, Loyola Marymount University · Scott James, GMI Engineering and Management Institute · Sundaresan Jayaraman, Georgia Institute of

Technology · Johndan Johnson-Eilola, Purdue University · Douglas Jones, George Washington University · John Kelly, Arizona State University · Robert Knighton, University of the Pacific · Bill Koffke, Villanova University · Celal N. Kostem, Lehigh University · George V. Krestas, De Anza College · Tom Kurfess, Carnegie Mellon University · Ronald E. Lacey, Texas A & M University · Frank Lee, Bellevue Community College · Gerald J. Lemay, University of Massachusetts-Dartmouth · Jonathan Leonard, Saginaw Valley State University · Gloria Lewis, Wayne State University · John Lilley, University of New Mexico · L. N. Long, Pennsylvania State University · Robert A. Lucas, Lehigh University · Sharon Luck, Pennsylvania State University · Daniel D. Ludwig, Virginia Polytechnic Institute and State University · Arthur B. Maccabe, University of New Mexico · Jack Mahaney, Mercer University · Bill Marcy, Texas Tech University · Diane Martin, George Washington University · Bruce Maylath, University of Memphis · John McDonald, Rensselaer Polytechnic Institute · Olugbenga Mejabi, Wayne State University · Steve Melsheimer, Clemson University · Craig Miller, Purdue University · Howard Miller, Virginia Western Community College · Andrew J. Milne, The Leonhard Center for Innovation and Enhancement of Engineering Education · Pradeep Misra, Wright State University · Charles Morgan, Virginia Military Institute · F. A. Mosillo, University of Illinois · Stanley Napper, Louisiana Technical University · John Nazemetz, Oklahoma State University · James Nelson, Louisiana Technical University · Brad Nickerson, University of New Brunswick · Col. Kip Nygren, West Point Military Academy · Adebisi Oladipupo, Hampton University · Joseph Olivieri, Lawrence Technological University · Joseph Olson, University of South Alabama · Kevin Parfitt, Pennsylvania State University · John E. Parsons, North Carolina State University · Steve Peterson, Lawrence Livermore National Lab · Larry D. Piper, Texas A & M University · Jean Landa Pytel, Pennsylvania State University · Mulchand S. Rathod, Wayne State University · Philip Regalbuto, Trident Technical College · Thomas Regan, University of Maryland-College Park · Larry G. Richards, University of Maryland, College Park · Larry Riddle, Agnes Scott College · Lee Rosenthal, Fairleigh Dickinson University · Isidro Rubi, University of Colorado at Boulder · Vernon Sater, Arizona State University · Dhushy Sathianathan, Pennsylvania State University · Kenneth N. Sawyers, Lehigh University · Carolyn J. C. Schauble, University of Colorado at Boulder · William Schiesser, Lehigh University · Bruce Schimming, Howard University · Keith Schleiffer, Battelle · Murari J. Shah, Technical Graphics · Howard Silver, Fairleigh Dickinson University · Ed Simms, University of Massachusetts · Thomas Skinner, Boston University · Gary Sobczak, Purdue University · Jim Strumpff, Mercer University · Habib Taouk, University of Pittsburgh-Titusville · Massoud Tavakoli, GMI Engineering and Management Institute · Greg Taylor, Northern Arizona University · Ronald Terry, Brigham Young University · Ron Thurgood, Utah State University · Stephen Titcomb, University of Vermont · Joseph Tront, Virginia Polytechnic Institute and State University · Israel Ureli, Ohio University · Bert Van Grondelle, Hudson Valley Community College · Lambert VanPoolen, Calvin College · Akula Venkatram, University of California at Riverside · Thomas Walker, Virginia Polytechnic Institute and State University · Richard Wilkins, University of Delaware · Frazier Williams, University of Nebraska · Billy Wood, University of Texas-Austin · Nigel Wright, University of Nottingham

10 Great Engineering Achievements

In celebration of its silver anniversary, the U.S. National Academy of Engineering identified the Ten Great Engineering Achievements accomplished during the organization's first 25 years. Initially selected because they represent major breakthroughs, these achievements have initiated whole new areas of engineering. In the pages that follow, we note the types of design problems faced by interdisciplinary teams of engineers who work in these fields, and we represent contemporary examples in the photographs. Many of the applications you will encounter throughout *The Engineer's Toolkit* are based on these achievements.

1

SPACE TRAVEL

Although 1972 marked the last in a series of moon landings begun in 1969, the Apollo mission laid a foundation for a whole new generation of space shuttle missions that were dedicated to gathering information about the universe and that continue to test the human ability to travel in space.

Design Problems Several key design problems had to be solved to support the Apollo mission to land humans on the moon. The spacecraft required a new inertial navigation system; the lunar lander ascent engine had to work perfectly because there was no backup engine; the spacesuits had to protect the astronauts in a hostile environment and yet be flexible enough to allow movement; and the Saturn V rocket, which powered all the Apollo flights, had to be 15 times more powerful than the biggest rockets available in the early 1960s.

Application Areas Today, rockets launch deep space probes such as the Voyager, which continues to send back images from as far away as Venus and beyond, while space shuttles launch information-gathering satellites and scientific equipment. The Hubble Space Telescope (HST) was launched from the

The space shuttle Endeavour floats above the earth at an altitude of 381 miles, with the west coast of Australia forming the backdrop for the 35mm frame. While perched on top of a foot restraint on the Endeavour's Remote Manipulator System arm, astronauts F. Story Musgrove (top) and Jeffrey A Hoffman wrap up the last of five space walks. They have succeeded in their mission to repair the Hubble Space Telescope.

space shuttle Discovery on April 25, 1990. NASA had designed HST to allow scientists to view the universe up to 10 times more sharply than they could with earthbound telescopes. Unfortunately, scientists soon discovered that the primary mirror in HST was flawed and could not focus properly. It was another space shuttle, the Endeavour, that operated as a kind of Hubble repair station.

Engineering Disciplines Although the fully operational Hubble Space Telescope now stands as an impressive

achievement for NASA, the repair job itself was perhaps an equally important achievement. Materials engineers helped develop metals, plastics, and other materials that could withstand the rigors of the launch and the environment in space. Mechanical engineers helped develop the mechanical structures that position HSTs mirrors and other moving apparatus and electrical engineers helped develop the complex computer, communication, and power systems.

2
APPLICATION SATELLITES

Application satellites and other spacecraft orbit the earth to capture, relay, and transmit specific types of information, or to perform manufacturing processes that rely on special properties of the extraterrestrial environment, such as zero gravity.

Application Areas Satellite systems provide information on weather systems, relay communication signals around the globe, survey the earth and outer space to map uncharted terrain and provide navigational information for vehicles on land, in the oceans, and in the air. The Endeavour has been involved in NASA's Mission to Planet Earth, which is designed to help the international scientific community better understand which environmental changes are caused by nature and which are induced by human activity. Throughout 1994 the shuttle orbited the earth with the Spaceborne Imaging Radar-C and X-Band Synthetic Aperture Radar system, which illuminates the earth with microwaves, allowing detailed observations at any time, regardless of weather or sunlight conditions. (See photo below.)

Design Problems A satellite or spacecraft launching system must be designed to generate enough thrust to escape the earth's atmosphere. Once free, it needs to maintain a stable orbit around the earth. In addition, the hull needs to be light and yet strong enough to withstand the stress of the liftoff.

Engineering Disciplines Space-based inventions, such as the Spaceborne Imaging Radar-C (SIR), and satellites in general are a result of the cooperative efforts of aerospace engineers who help develop the systems that put satellites in space, and of chemical, mechanical, and electrical engineers who assist in the development of the radar and imaging systems for applications such as SIR.

3
MICROPROCESSORS

A microprocessor is a tiny computer, smaller than your fingernail, that combines the control, arithmetic, and logic functions of large digital computers.

Application Areas With its small size and powerful capabilities, the

A technician working on the Bit Serial Optical Computer (BSOC), an optical computer that stores and manipulates data and instructions as pulses of light. To enable this, the designers (Harry Jordan and Vincent Heuring at the University of Colorado) developed bit-serial architecture. Each binary digit is represented by a pulse of infrared laser light 4 meters long. The pulses circulate sequentially through a tightly wound 4-kilometer loop of optical fiber some 50,000 times per second. Other laser beams operate lithium niobate optical switches which perform data processing.

applications of microprocessors range from operating remote television controllers or VCR recorders to providing the computational power in hand-held calculators or personal computers. Microprocessors can also be found in communication devices, such as networks that connect computers around the globe, and in automobiles, ships, and airplanes.

Design Problems Key design problems involved in creating microprocessors include miniaturization, increasing speed while controlling the heat produced, and searching for materials stable and reliable enough to store, process, and transmit data at high speeds. For decades engineers have improved the performance of computers by increasing the number of functions contained on the CPU chip. Ultimately this approach created a bottleneck in switching between these functions. In the early 1980s, chip designers addressed this issue and developed a concept of Reduced Instruction Set Computing (RISC) which improves efficiency through the high-

This image of the area around Mount Pinatubo in the Philippines was acquired by the Spaceborne Imaging Radar-C and X-Band Synthetic Aperture Radar system aboard the space shuttle Endeavour in April 1994. This false color image shows the main volcanic crater on Mount Pinatubo produced by the June 1991 eruptions and the steep slopes on the upper flanks of the volcano. The red color shows the rougher ash deposited during the eruption. The dark drainages are the smooth mudflows that continue to flood the river valleys after heavy rains. This radar image helps identify the areas flooded by mudflows, which are difficult to distinguish visually, and assess the rate at which the erosion and deposition continue.

speed composition of an optimized minimal set of instructions. Research and development is on-going in another area of microprocessor design as well: optical computing. A dream since the 1940s, optical computing represents a fundamental change in how switching occurs—through optical signals rather than electronic signals. Since a computer is nothing more than a complex system of on/off switches, the speed at which the switches can turn off and on is the single most critical factor in determining the computer's performance. Engineers can design optical switches that operate well into gigahertz range, while electronic switches are currently restricted to about 100 megahertz.

Engineering Disciplines An application such as optical computing is a result of many years of collaboration between electrical engineers who

Inspection of the largest, most powerful energy-efficient jet engine, which was designed for the Boeing 777 jets.

design the complex laser systems, computer engineers who work on the computational structures, and chemical engineers who develop the actual lasers.

4

JUMBO JET

Much of the success of jumbo jets (747, DC-10, L-1011) can be attributed to high-bypass fanjet engines, which allow the planes to fly farther using less fuel and with less noise than previous jet

engines. Jumbo jets also have an increased emphasis on safety. For example, a 747 has four main landing-gear legs instead of two; a middle spar was added to the wings in the event one is damaged; and redundant hydraulic systems operate the critical system of elevators, stabilizers, and flaps that control the motion of the plane.

The newest jumbo jet is Boeing's 777, which was entirely developed using computer-aided design systems.

Application Areas The technological advances of superior fuels, engines, and materials achieved during the creation of the jumbo jet benefited the space program and the military.

Design Problems A major design problem faced by creators of the jumbo jet was creating an engine and fuel supply that could generate sufficient horsepower and thrust needed to lift the huge aircraft off the ground. The hull had to be strong enough to withstand the stress of the flight, without being so heavy that fuel efficiency was lost. The internal environment had to be comfortable for the passengers, especially during liftoff and landing.

Engineering Disciplines Engineers have been developing increasingly powerful jet engines ever since Englishman Sir Frank Whittle developed the first jet engine prototypes in 1937. The evolution of Whittle's engine into the Boeing 777 engine was largely due to the efforts of mechanical engineers who specialized in dynamics, thermodynamics, combustion systems, and materials.

5

MEDICAL IMAGING

A CAT (computerized axial tomography) scanner is a machine that generates three-dimensional images or slices of an object using x-rays. A series of x-rays is generated from many angles, encircling the object or patient. Each x-ray measures a density at its angle, and by combining these density measurements using sophisticated computer algorithms, an image can be recon-

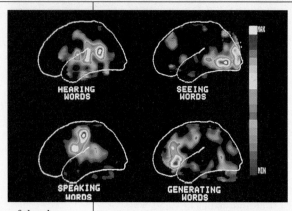

The PET scans shown here reveal localized brain activity under four different conditions, all related to language. Physicians use the PET scanner to diagnose brain and heart disorders and certain types of cancer.

structed that gives clear, detailed pictures of the inside of the object. A PET (positron-emission tomography) scanner reveals locations of intense chemical activity within the body. Sugar labeled with radioactive isotopes that emit particles called positrons is first injected into a person's bloodstream. These positrons collide with electrons made available by chemical reactions in the body. The scanner detects the energy released by these collisions and maps metabolic "hot spots"—regions of an organ that are most chemically active at the time.

Application Areas Doctors use diagnostic radiology to both detect and diagnose diseases, and in combination with other medical techniques, to treat diseases. These devices can also be used to explore the structure of both organic and inorganic materials.

Design Problems The key design problem lies in getting a clear picture without harming the patient or technician. In addition, the machine needs to withstand the various fields it produces.

Engineering Disciplines The PET and similar medical scanning systems are a product of biomedical engineering, a very specialized field of engineering that combines electrical engineering with medicine. Also involved are materials engineers—mechanical engineers who develop the special structures that contain, direct, and withstand the sometimes hazardous electromagnetic and radioactive emissions often used in such systems.

6

ADVANCED COMPOSITE MATERIALS

A composite consists of a matrix of one material that has been reinforced by the fibers or particles of another material. The choice of the composite is determined by the need of the application. For example, does the application require a material that is strong, flexible, stiff, lightweight, heavy, heat-tolerant, porous, dense, or wear-resistant?

Application Areas Advanced composites can be found in most products, including automobile components, communication systems, building materials, artificial joints and organs, and machine parts.

Design Problems Composite material designers must determine how the various materials will interact with each other and how the composites will act over time given the often high-stress situations in which they are used.

Engineering Disciplines Mechanical engineers are most frequently involved in the development of composite materials. Working with chemists and sometimes chemical engineers, mechanical engineers design applications, such as artificial limbs, for particular composite materials.

7

CAD/CAM/CAE

CAD (computer-aided design) refers to computer systems used by engineers to model their designs. These systems may include plotters, computer graphics displays, and 2-D and 3-D modelers. CAM (computer-aided manufacturing) systems are used to control the machinery or industrial robots used in manufacturing the parts, assembling components, and moving them to the desired locations. CAE (computer-aided engineering) systems support conceptual design by synthesizing alternative prototypes using rule-based systems; they also support design verification through rule checking of CAD models. CAD/CAM/CAE systems are intended to increase productivity by optimizing design and production steps, and by increasing flexibility and efficiency.

Application Areas CAD/CAM/CAE systems are used in all engineering design disciplines to support product design, test, and production.

Design Problems Engineers must correctly convert real-world specifications to valid computer models, design appropriate computer tests for the model, and design computer systems to accurately and economically manufacture the final product.

Engineering Disciplines CAM has had a major effect on the work of many industrial engineers. Switching over from labor-intensive manufacturing to computer-aided manufacturing has changed and will continue to change the national and international industrial environ-

Zina Bethune, ballet teacher, has special long artificial hip implants coated with cobalt and chrome. She needed the implants because she suffers from degenerative arthritis.

This automated integrated circuit insertion device, designed by Lockheed, is installing an integrated circuit into a circuit board for a satellite program.

ment. Because they use CAM systems themselves, industrial engineers are heavily involved in their development. Computer and electrical engineers design the computer and control systems and write the software that drive CAM systems.

8

LASERS

Light waves from a laser (that is, light amplification of electromagnetic waves by stimulated emission of radiation) have the same frequency and thus create a beam with one characteristic color. More importantly, the light waves travel in phase, forming a narrow beam that can easily be directed and focused.

Application Areas CO_2 lasers can be used to drill holes in materials such as ceramics, composite materials, and rubber. Medical uses of lasers include repairing detached retinas, sealing leaky blood vessels, vaporizing brain tumors, removing warts and cysts, and performing delicate inner-ear surgery. Lasers are used in scanning devices to scan Universal Product Codes. High-power lasers also are used in weapons and to create 3-D holograms.

Design Problems A key concern is controlling the heat and power generated by lasers so the laser does not harm the patient or the product.

Engineering Disciplines Although they were not heavily involved in the original design and development of lasers in the 1950s, engineers have developed applications for them in

Automatic analysis of DNA is performed with laser beams by Leroy Hood and Jane Sanders, biologists at California Institute of Technology. This DNA sequenator is also called the "Gene Machine."

many areas: Electrical engineers use them in optical fiber communications, civil engineers use them to perform accurate surveying, and mechanical engineers use them for precise cutting of metal parts.

9

FIBER OPTIC COMMUNICATIONS

An optical fiber is a transparent thread of glass or other optically transparent material. It can carry more information than either radio waves or electrical waves in copper telephone wires. In addition, fiber optic communication signals do not produce electromagnetic waves that cause cross-talk noise on communication lines. The first transoceanic fiber optic cable was laid in 1988 across the Atlantic. It contains four fibers that, together, can handle up to 40,000 calls at one time.

Application Areas An increasing number of communications and computer systems are converting to fiber optics due to its enormous information capacity, small size, light weight, and freedom from interference.

Design Problems A key design problem is to create cables that can withstand the stress of being buried under ground or under the ocean. Another design problem is the need to

keep the various signals independent and free from outside interference.

Engineering Disciplines Mechanical engineers design the manufacturing systems that produce glass and plastic optical fibers. Electrical engineers design the transmitters, amplifiers, and receivers that, along with the fibers, carry the optical signals from source to destination.

10

GENETIC ENGINEERING

A genetically engineered product is created by splicing a gene that produces a valuable substance from one organism and placing it into another organism that will multiply itself and the foreign gene along with it. Genes are artificially recombined in a test tube, inserted into a virus or bacteria, and then inserted into a host organism in which they can multiply. Once the new organism has been created, a system has to be designed to produce and process the product in large quantities at a reasonable cost.

The first commercial product of genetic engineering was human insulin, which appeared commercially under the trade name Humulin. The molecules are produced by the genetically engineered bacteria and are then crystallized into human insulin.

Application Areas In addition to creating new drugs and vaccines, genetic engineering has been used to create bacteria that can clean up oil

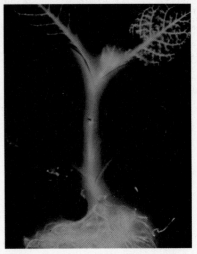

Researchers incorporated firefly gene codes for the enzyme that catalyzes the chemical reaction to release energy in the form of light, into the DNA of a tobacco plant.

spills and detoxify wastes. The process is also used to create genetically altered plants that are pest and disease resistant or have certain desirable characteristics such as improved taste, shipping hardiness, or longer shelf life.

Design Problems Engineers must translate the laboratory work of biologists into the large-scale commercial manufacturing systems that are robust, safe, and cost effective.

Engineering Disciplines Mechanical engineers design equipment for growing large quantities of genetically engineered organisms, chemical engineers design processes for separating out the desired end product, and environmental engineers evaluate the potential impact on the environment.

Thinner than a human hair, an optical fiber can carry more information than conventional radio waves or electrical waves.

Exploring Engineering

Joe King
Department of Electrical Engineering
University of the Pacific, Stockton, California

 Addison-Wesley Publishing Company, Inc.

Reading, Massachusetts · Menlo Park, California
New York · Don Mills, Ontario · Harlow, U.K. · Amsterdam
Bonn · Paris · Milan · Madrid · Sydney · Singapore · Tokyo
Seoul · Taipai · Mexico City · San Juan, Puerto Rico

This is a module in *The Engineer's Toolkit*, an
Addison-Wesley SELECT edition. Contact your sales
representative for more information.

The Engineer's Toolkit and SELECT are trademarks of
Addison-Wesley Publishing Company, Inc.

Photo Credits:

Chapter 1: © R. Ian Lloyd/Westlight.
Chapter 2: Courtesy of EG&G Structural Kinematics.
Chapter 3: Courtesy of The Boeing Company.
Chapter 4: © David Young-Wolff/Photo Edit.
Chapter 5: © Paul Chesley/Tony Stone Images.

Figure Credits:

Figures 1-2, 1-3, and 3-1:
Courtesy of the University of California, Lawrence
Livermore National Laboratory, and the Department
of Energy, under whose auspices the work was performed.

ISBN: 0-8053-6354-8

Addison-Wesley Publishing Company, Inc.
2725 Sand Hill Road
Menlo Park, CA 94025

Contents

Introduction

This module is designed to help you make the most of your college education and to help prepare you for an engineering career. Through descriptions of actual engineering projects in Chapters 1 and 3, you will learn about the stages of the engineering design process and the skills you will need to succeed as a practicing engineer. Other chapters describe the various engineering disciplines and the wide range of career possibilities available to you upon graduation. Throughout the module you will be given practical advice on everything from how to interview for a job to how to stay current in your field once you have entered the job market. Exercises at the end of each chapter are designed to help you explore further the material presented in the chapter.

To give you insight into the "real world" of engineering, the author conducted interviews with 13 practicing engineers in a variety of fields. Excerpts from these interviews can be found at the start of each chapter and are interspersed throughout the module. The full text of the interviews can be found on our World Wide Web site, http://www.aw.com/cseng/toolkit/.

1 What Does an Engineer Do?

"A couple of months after I joined this group, I hooked up with the first Alpha chip project," says Sharon Britton, a Principal Hardware Engineer for Digital Equipment Corporation (DEC). "That was an exciting project. We were really breaking new ground in terms of circuit design and logic design. At the time there was another chip being designed with the same process technology, but we decreased the clock cycle by a factor of two and a half times. Our most important challenge was, 'How do you make something run that fast?'"

Each new project has brought new challenges for Sharon Britton and her colleagues in DEC's Digital Semiconductor Engineering Group. "I'm on my third Alpha project now, and they grow logically more and more complex," says Britton. "The challenge is to look at the architectural problems and figure out a way to redefine the problem, or to squeeze it into a very simple model of logic. And there are a slew of other issues involved, such as how to maintain reliability when you're pushing that much current through the wires.

"Another problem we have to deal with is power consumption," she adds. "So we're looking at fitting everything into a given die size and dealing with complex logic; and we're also trying to manage the power. Sometimes you feel boxed in by all the constraints."

Designing and developing new computer circuits is a team effort, Britton notes. "When you include the verification engineers, the layout designers, the architects, and the implementors, you get up to well over 100 people on a project. The project structure, however, is such that the chip is broken up into six 'boxes,' such as memory management and the instruction execution unit. So even though the project is large, you end up working with a fairly small team of people—probably less than ten on a daily basis."

Every computer chip project moves through three distinct phases, according to Britton. "The first part, when you're working with the architects and you're defining feasibility, is very creative and challenging, as you're thinking about all of these high-level problems. The next phase of the project is the actual circuit design, which includes a considerable amount of detailed SPICE simulation and analysis. Finally, at the back end, everyone is working on checking the layout, checking the reliability, doing the physical verification."

"These phases are very different and require very different skills," she adds. "I wouldn't want to stay in any one phase forever. But by the time you finish the back end of one project, you're ready to go to the front end of the next project. It's fun. It keeps the job interesting."

INTRODUCTION

This chapter answers the question, "What does an engineer do?" The best way to answer this question is to describe an actual engineering project. This chapter takes you from the beginning to the end of a real project, in which a young electrical engineer designs a data recorder. Although most engineering projects are complex and require the skills of teams of engineers, for the sake of simplicity this chapter presents a relatively straightforward project. During the course of this or any other project, the engineer performs many tasks; however, our primary focus is on design.

1-1   ENGINEERING AND THE DESIGN PROCESS

A typical engineering project begins with a problem or an idea for a new or improved product. The design of the problem solution or new product or process then goes through a series of stages. The five-step design process parallels a general procedure that can be applied to solving any problem.

General Problem-Solving Procedure	Five-Step Design Process
1. Define the problem.	Develop functional specifications.
2. Gather information.	Develop the concept design.
3. Generate and evaluate potential solutions.	Generate design alternatives.
4. Refine and implement a solution.	Select and model the best alternative.
5. Test and verify the solution.	Test and verify the design.

During the first stage of an engineering project, the engineers, management, clients, and others involved in the project develop functional specifications. Much analysis and discussion takes place and the market is studied before the project proceeds. During the concept design stage, the engineers gather as much information as possible about solutions to the design problem and develop a basic approach to the problem solution. During the third and fourth stages, the engineers generate alternative designs and then select and model the best among the alternatives. The model may be computer-based, a physical prototype, or a combination of the two. In any case the model is tested during the fifth and final design stage to verify that it meets the functional specifications. Once the design is verified, the product is manufactured and marketed.

Throughout the process it is critical that complete and accurate documentation be maintained. Documentation can include personal daily notes, meeting minutes, monthly progress reports, and final project reports.

As indicated by the arrows in Figure 1-1, the five-step design process is iterative. After discussions with engineers or other specialists, a client may decide to cancel the project. Sometimes, while gathering information, the engineers determine that the project cannot proceed as originally proposed, and they must modify the functional specifications. The best design alternative may require parts or materials that are not presently available or that are judged to be too expensive. During testing, engineers occasionally find that the supposed best design was not best after all, and so they return to one of the alternatives. Even during the manufacturing stage, problems can arise that require a redesign. At any point in the design process described in this chapter, problems can occur that force the engineers to return to an earlier design stage and try another approach.

The design process is often not only iterative but also concurrent. That is, once the specification stage is completed, as many activities as possible occur simultaneously, to speed up the design process and decrease the time to market. Large projects are broken down into components that can be designed separately. Sometimes single engineers design the components; more often small teams of engineers design them. For example, while electrical engineers design a video board, industrial engineers design the layout of the assembly line where the board will be assembled, drafters create schematic drawings, and technicians lay out the printed circuit boards.

FOCUS ON

WHY I CHOSE ENGINEERING

"I chose engineering for two reasons. One, I enjoyed math and science as a boy. As I grew older, I found that I enjoyed science more, and in particular, physics. I could have been a physicist because I liked the subject so much. But I didn't think I could make a living at physics in the private sector, and I didn't want to spend the rest of my life on a campus, teaching. Engineering allowed me to do a lot of things I enjoy in physics."

Joe Engel
Structural Engineer
Engel & Company
Engineers

"When I was in high school, I considered two career paths, engineering and culinary arts. My parents made a big impression on my career decision. When I was going through high school, I excelled in math. So my parents said 'If you're going to try something and go into it, you can always change your mind. So you probably want to start in the engineering area first, because it aligns closer to what you've already learned and it will stay fresher.' As I look back, I can say it was the right decision. I enjoy it a lot."

Greg Raco
Project Manager
Ocean Spray, Inc.

"When I was a kid growing up, I was always interested in solving problems. I enjoyed challenges, which is a big part of engineering, I think. And my dad was an architect. I used to go down to his office. I was always impressed by the guys in the white shirts, leaning over their drafting boards."

Ian Buist
Vice-President
S. L. Ross Environmental
Research Ltd.

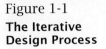

Figure 1-1
**The Iterative
Design Process**

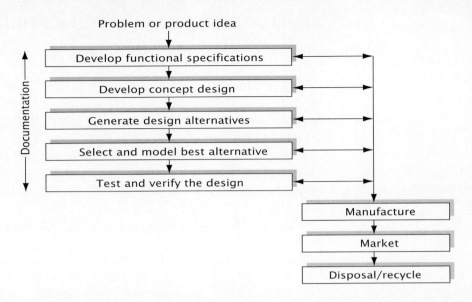

Project concurrency also implies involvement of the sales, marketing, and manufacturing departments in the design process. Sales engineers usually know best what customers want. Marketing has to advertise and promote the product. Manufacturing must be involved to ensure that the ultimate product can be made. For example, the electrical engineer who designs a computer's motherboard works closely with the manufacturing engineer who will design the production line that will mount electrical components in the motherboard. The mechanical engineer who designs an airplane's landing gear cooperates with the industrial engineer who will design the layout of the factory that will manufacture the gear. The civil engineer who designs an office building assists the construction engineer who puts together a team of subcontractors that will be responsible for the building's construction. Design engineers should have some under-standing of the manufacturing or construction process so they can work more easily with those involved in these efforts.

Unfortunately, it is difficult to design a manufacturing process without a complete design of the product to be manufactured. Therefore, it is nor-mal for the manufacture stage to begin only after the engineers have selected the best alternative design, proceeding simultaneously with the test and verification stage. Increasingly, however, design and manufactur-ing engineers are being encouraged to cooperate earlier in the design pro-cess, so their jobs are more concurrent and they avoid designing a product that cannot be manufactured.

An engineer must consider not only the functional specifications and manufacturability of a design but also its complete life cycle. The life cycle of a design includes such issues as maintenance costs, power effi-ciency, long-term product safety, and disposal or recycling issues. The issue of product disposal or recycling involves not only monetary but also environmental costs. Chapter 5 discusses environmental impact issues.

1-2 DEVELOP FUNCTIONAL SPECIFICATIONS

Whatever the problem or proposed idea, a design must be clearly defined before work can begin on it. Engineers, managers, clients, and technicians must agree on exactly what the problem is and how they expect the design to solve the problem. These expectations are then translated into and quantified as the functional specifications. Budgets and work schedules must also be developed. In some instances functional specifications, budgets, and schedules combine to act as a contract between those who pay for the problem solution and those who are paid to solve the problem. A large or complex project may have so many functional specifications that a 100-page document is required to list them.

The functional specifications can include many components, depending on the type of project. They may include requirements for documentation, weight, size, safety, testing, intended users, reliability, maintenance, materials used, quantity produced, ergonomics, aesthetics, shelf life, packaging, product life, and meeting engineering, legal, and environmental standards. Engineers often create further specifications as a result of the design process. For example, they may specify certain size and weight restrictions of parts of a design to ensure that the complete design meets its overall functional specifications.

Case Study: The Data Recorder Project

The data recorder project began when a customer, a team of three physicists, came to a research facility with a problem. They needed to know the forces to which a projectile was subjected when passing through various materials, such as steel and concrete. Several management-level engineers at the research facility met with the physicists. The managers decided their engineers could design a recording device that would

- Be small enough to fit within a projectile
- Record the forces to which the projectile was subjected
- Survive those forces
- Transmit its recorded data to an external personal computer

The physicists required the recorder to have two accelerometers to measure, respectively, the longitudinal and transverse acceleration of the projectile. Using Newton's Law, $F = ma$, the forces acting on the projectile could be computed. The recorder would detect the collision with the target and then collect 200,000 samples of acceleration data per second per channel for 10.2 milliseconds. To eliminate noise interference, each accelerometer output would be filtered to remove all frequencies above 20 kHz.

The project presented two main problems. First, the physicists estimated that the acceleration would reach a maximum of −20,000 g when the projectile impacted its target. The recorder had to survive this impact for its data to be retrieved. The second problem was that the recorder could be no larger than roughly the size of two fists, the space available for it within the projectile.

After several meetings the physicists and engineers agreed on the set of requirements listed in Table 1-1.

Table 1-1 Recorder Specifications

Feature	Specification
Channels	Two, longitudinal and transverse
Transducer type	Piezoelectric accelerometers
Sample rate	200,000 samples per second per channel
Data collection	10.2 milliseconds
Bandwidth	0 to 20 kHz
Range	0 to −20,000 g
Resolution and accuracy	0.8% and 5%, respectively
Triggering	Must detect collision with target
Power supply	Battery pack
Size	7.1 cm external diameter, 15.2 cm long
Data recovery	Interface to external PC

Engineers rarely receive a set of performance requirements and are told, "Spend all the money you want, and take all the time you want." Usually they must adhere to a budget and a schedule. The physicists and engineers on the data recorder project agreed on the cost and time specifications listed in Tables 1-2 and 1-3, respectively.

Table 1-2 Recorder Budget

	Hardware	Personnel	Total
Phase I - Research	$18K	$100K	$118K
Phase II - Design	$14K	$107K	$121K
Phase III - Development	$20K	$158K	$178K
Phase IV - Testing	$6K	$25K	$31K
TOTAL	$58K	$390K	$448K

Table 1-3 Recorder Project Schedule

Phase I - Research	July to December
Phase II - Design	December to April
Phase III - Development	April to November
Phase IV - Testing	October to December
TOTAL	18 months

1-3   DEVELOP THE CONCEPT DESIGN

Once the specifications are established, they are given to one or more engineers. The engineers begin their design work by gathering information and researching possible solutions. From this research they develop a concept design. Typically engineers generate a number of potential concepts, from which they select the best one using a process of research and testing. Sometimes engineers must repeat the process because, during later design stages, they may discover that the concept design was not the best after all.

The project budget and schedule act as constraints on the amount of time engineers can spend gathering information and developing the concept design. The budget includes the wages of the engineers, so it may be the final arbiter of how much time they can spend gathering information.

Staying on schedule is a major frustration in most engineering projects. Many projects include contracts that stipulate significant monetary penalties for going beyond the allotted time.

While developing the concept design, engineers must also consider the number of units to be built. If they are designing a fuel injection unit that will be reproduced in 100,000 automobiles, the engineers might be justified in adding an extra week to the schedule to save one dollar in manufacturing costs. On the other hand, if only one unit is to be built, the engineers should keep the amount of time spent on the concept design to a minimum. This helps reduce the cost of the project. Even though the cost of building the unit can be greater, the total cost of design and development is less. In all projects, engineers must strike an appropriate balance between the cost of the design and development and the cost of construction or manufacture.

The Data Recorder Concept Design

The research facility assigned the recorder project to Elizabeth. It was her second project since receiving her engineering degree. An experienced technician, Ralph, was assigned to the project to help her gather information and generate possible approaches to the problem.

The task of gathering information began as Elizabeth and Ralph studied the functional specifications (see Table 1-1) and what they implied. For example, the specification to sample the incoming analog acceleration signal 200,000 times per second required the use of high-speed circuitry and design techniques. Combined with the sample rate, the 10.2-millisecond sample time established the amount of memory required: 2kilobytes per channel. The 20 kHz cut-off frequency implied that a low-pass filter was required at the input to each data collection channel. These filters removed unwanted high-frequency noise that resulted from vibrations of the projectile.

Although Elizabeth had studied microprocessors in school, she had never designed a microprocessor-based system before. So she began the project by spending two weeks studying books and journals on the subject. Once she had a good handle on the microprocessor applications, she and Ralph, who had previously assisted an engineer on a microprocessor-based design, developed the block diagram shown in Figure 1-2.

The block diagram represents a concept design, or general approach to the problem. Note that it includes two channels, for recording both longitudinal and transverse forces, as well as memories for storing the recorded data and an external interface circuit for transferring the recorded information to an external personal computer. A concept design for a civil engineering project might include sketches of a structure, and a concept design for a mechanical engineering project might include sketches of a machine or engine.

Figure 1-2
Concept Design for the Recorder Project

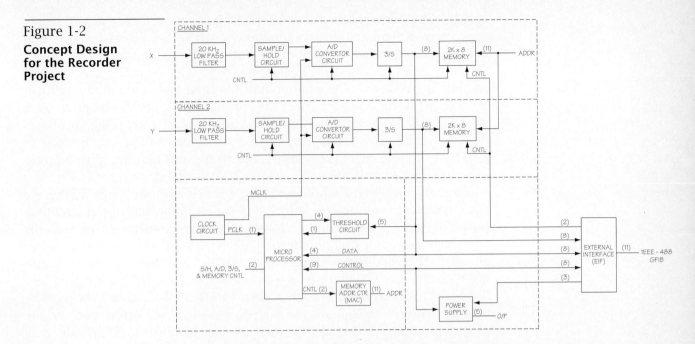

The managers of the recorder project told Elizabeth that only 50 to 100 recorders would be manufactured, so manufacturing costs were not a significant factor. Nonetheless, she sought low-cost concept designs, without devoting an excessive amount of time to the search. As much as possible, she used readily available, low-cost parts and designed simple circuits. Simple circuits would not only cost less but also consume less space and tend to be more reliable.

Typically, once engineers settle on a concept design, they must get the approval of management before proceeding to the next design stage: generating specific design alternatives. Elizabeth presented her reasons for selecting her concept design in a project review presentation before the customer (the physicists, another engineer, and two managers). At the presentation, Elizabeth displayed a transparency of her block diagram and described what each circuit would do and how the circuits would work together. Those who attended asked questions, forcing Elizabeth to defend her ideas. They also asked her to explain her reasons for rejecting other possible concept designs. The experience was stressful, but Elizabeth decided to learn as much as she could from it since she would have to repeat the process every few months throughout the 18-month project.

1-4 GENERATE DESIGN ALTERNATIVES

Once engineers have developed a concept design, they are ready to start generating specific design alternatives. In this phase engineers consider not only the functional specifications but also manufacturability, testability, marketing, field support, and disposal or recycling. Elizabeth relied on Ralph's experience in manufacturing. Since her recorder would be used only by the physicists who funded the project, it would not be marketed. Field support and disposal were also not big issues, given the small number of recorders that would be manufactured.

After Elizabeth successfully defended her concept design, she began generating potential solutions for the circuits in her block diagram, and Ralph began to design the recorder's power supply. Elizabeth started with the filter circuit and easily found the design she wanted in an electrical engineering textbook. Once Ralph built the filter she chose, however, testing showed it did not quite meet the specifications. Elizabeth had to alter its design. She did so using mathematical analysis, combining two types of filters. Interestingly, she developed a new type of filter that worked so well that the research facility asked her to present the design to a group of engineers and physicists.

Elizabeth designed each circuit in the block diagram sequentially. As she completed each design, Ralph implemented and tested the circuit in the laboratory. Testing usually showed that some modification of each circuit was required or that an alternative design would have to be pursued. Elizabeth designed the microprocessor circuit last. While Ralph was building it in the lab, Elizabeth wrote its software. Together they tested and debugged the complete system shown in Figure 1-2. About six months into the project, Elizabeth and Ralph, ahead of schedule, had all the recorder circuits working together in the laboratory. Elizabeth gave another project review presentation. This time, she had more them a dozen transparencies to illustrate her 30-minute talk.

1-5   SELECT AND MODEL THE BEST ALTERNATIVE

Engineers almost always have several alternatives from which to choose when designing a system. They must rely on their education and experience and, if possible, some testing to help them make the most appropriate choice in each case.

Elizabeth chose her initial filter design from many available designs. Her final choice was a unique design of her own. She had many commer-

---

FOCUS ON

WHY I CHOSE ENGINEERING

"When I got out of high school, I first went to pharmacy school, mainly because I was good at math and science. At the time, high school girls were not encouraged to become engineers, so I never even considered it.

While at pharmacy school, I took several computer programming electives. After thinking more about what I wanted to do, I dropped out of pharmacy school for a while, took a couple of engineering electives, decided I wanted to be a designer, and finally transferred into engineering school."

> Sharon Britton
> Principal Hardware Engineer
> Digital Equipment Corporation

"Since I can remember I always liked mathematics and sciences in general. In high school I learned that engineering had to do with applying math and science to the design and building of things, so that's how I got interested in engineering.

While I was in my senior year of high school, I got interested in amateur radio. I bought a CB radio and I did the antenna installation myself. I started to study and learn about electronics on my own, and that is how I decided that I wanted to study electrical engineering."

> Jose E. Hernandez
> Electrical Engineer
> Lawrence Livermore National Laboratory

cially available microprocessors from which to choose. However, the sampling-speed requirement eliminated many of her choices, and space restrictions within the recorder eliminated most of the others. She settled on a small, fast processor from a reliable supplier. Her choice used more power than all of her other circuits combined, so Elizabeth designed a circuit that provided power to the microprocessor only when it detected impact with the target. This approach was necessary because of the limited power supplied by the recorder's battery.

There were many available memory chips Elizabeth could have used to store the recorder's data. As with the other components, she based her choice on speed, size, and power consumption.

The interface that would transfer the recorder data to an external personal computer presented a special problem. The personal computers used by the research laboratory and the clients had a specialized IEEE-488 GPIB port, in addition to the usual serial and parallel ports. This specialized port adheres to an interface standard specified by the Institute of Electrical and Electronic Engineers (IEEE). The clients preferred that the recorder use the IEEE interface, so Elizabeth had to design a special circuit that translated the data stored in the recorder into signals that the IEEE-488 GPIB port could then pass on to the personal computer memory.

It took about a month of testing and debugging for Elizabeth and Ralph to get the recorder circuits to work individually and then together as a complete system in the laboratory. Several circuits that worked fine alone had to be redesigned to work with the other circuits. The software did not work as originally written and required significant debugging. Finally, Elizabeth and Ralph felt they had the best design alternative. They were ready to model the design so they could fully test it before sending it to manufacturing.

1-6 TEST AND VERIFY THE DESIGN

The purpose of the test stage is to verify the solution, that is, to prove that it meets the functional specifications. First Elizabeth had to prove that her laboratory-based system met the appropriate specifications. She was then ready to move to the test stage, in which she and Ralph had to

FOCUS ON

WHY I CHOSE ENGINEERING

"I remember wanting to do things with computers even way back in junior high school. I remember trying to convince my shop teacher to get the entire class to help me wind electromagnets so we could put together an electromagnetic computer. This was back in the mid-1960s. Lo and behold, I convinced him; we made all the electro-magnets, and I made a computer. So I was always fascinated with computers even before they were really cool.

Knowing sort of what I wanted to do, I decided to attend the United States Air Force Academy, where I got my undergraduate degree in computer science. The reason I chose to go there was they had a very good under-graduate computer science program. Plus being in the Air Force gave me a tremendous opportunity to work with some very complicated computer systems after graduation."

Grady Booch
Chief Scientist
Rational Software
Corporation

prove the recorder functioned as specified within a projectile. Finally she had to prove it would work when subjected to –20,000 g.

Toward this end Elizabeth and Ralph developed a prototype recorder. A prototype is a model or first implementation of a design that is built specifically for verifying a design. This is particularly important when manufacturing production models will be expensive. For example, whenever an automaker plans to produce a new automobile, several prototypes are first constructed. During the test stage the prototypes are driven many thousands of miles and are subjected to every conceivable severe condition. Bugs in the design are corrected. Only then does the organization go into full production. Such thorough testing of prototypes can save millions of dollars on recalled autos. Even software designers create prototypes called alpha and beta software versions, which are fully tested by selected end users, before actually marketing the software.

The electronic components Elizabeth used to test her circuit designs in the laboratory were standard commercial integrated circuits. Since these parts would not survive the rigors of a 20,000 g deceleration if mounted on normal circuit boards, Ralph had to glue the circuits directly to specially designed circuit boards in the recorder. After procuring the necessary materials, he designed and directed the production of the printed circuit boards and mounted his power supply and Elizabeth's circuits on them.

Acquisition of the materials for the construction of an engineer's design is an important part of every project. Sometimes inexperienced engineers learn the hard way not to wait too long to start this task. They assume the parts will be there when they need them and they will have no problem acquiring them. Waiting for parts or materials to arrive can make a project fall behind schedule, so engineers should determine early in the project what parts are needed, get them ordered as soon as possible, and give special attention to parts that are rare or in great demand. Engineers should also order extra parts to replace failed ones.

A mechanical engineer designed the aluminum case shown in Figure 1-3, in which the recorder circuit boards were placed. The case was filled with a potting liquid that solidified to hold the boards and circuit components in place as the recorder was subjected to severe forces. In a laboratory, the case containing the potted recorder system was slammed into a hardened stainless steel surface, subjecting it to decelerations of up to 20,000 g. After each test, the recorder successfully passed its data to a personal computer via its external interface. The project was a success. No redesign was required.

As an electrical engineer, Elizabeth could build a prototype to test her designs before going into production. Engineers in some other fields are not so fortunate. A complete highway or office building can hardly be tested in the laboratory. However, parts of them can be. For example, highway and building construction materials are tested before being used in a project.

Many engineering projects test computer-based models, eliminating the need to build a prototype. This can be particularly cost effective for the design of a complex and expensive product, such as an automobile or airliner. As computers become less expensive and more powerful, computer-based modeling is becoming increasingly popular on both small and large projects. When neither a prototype nor a computer model can be created, engineers must rely on their own experience and the experience of other engineers.

Figure 1-3

The Case that Held the Recorder Within the Projectile

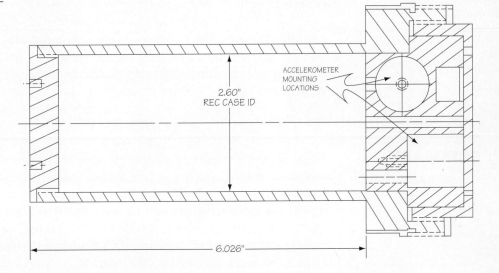

1-7   DOCUMENTATION

Engineers must maintain several types of documentation throughout every project. The most important document for any engineer is a complete and accurate project journal, which includes project notes and a daily record of project activities. The journal should be hardbound so pages cannot be removed, and copies of all project reports and memos should be included.

Whenever engineers start a new project, they should record notes taken during the initial project meetings with managers, other engineers, and technicians. This will get the project started on the right track. To stay on track, engineers must take complete notes at all subsequent meetings. Copies of all block diagrams and other design sketches should be part of the journal. Promises made by part suppliers, technicians, and managers should be recorded. Generally engineers must record everything of significance in their journals and date everything they record.

A project journal is useful in a number of ways. For example, most organizations require a written monthly progress report. A good record of all the activities of the previous month can ease the writing of this report. If an engineer has to call a supplier to find out where parts are, he or she can say, "According to my records, on March 2 John Smith at your company promised me that my parts would be here yesterday." A journal can also help support the originality of their ideas, as in a patent application or a patent defense.

After completing a project and verifying its solution, an engineer must write a final project report. This document will guide engineers who must update or revise the design sometime in the future. Most designs are later modified; therefore, the final project report must be clear and complete. It should list the functional specifications and describe every component of the design. It should also include explanations of all design decisions. If the project included any software, the final report should describe that too. Test results must also be documented to prove that specifications were met. Engineers may also be required to write a user's manual that describes how to use and maintain the device or machine they have designed.

Elizabeth spent several weeks writing the final documentation for her recorder project. She wrote a 4-page user's manual and a 50-page project report. In the report she included 12 pages of test data that proved her recorder met the functional specifications. Then she moved on to her next project.

SUMMARY

This chapter described a simple engineering project. Elizabeth, the project engineer, started with the functional specifications for a data recorder, from which she developed a concept design. The concept design described a global approach to the problem. With the help of her assistant, Ralph, she designed, implemented, tested, and debugged each circuit. Once she had all the circuits working individually, Elizabeth got the circuits to work in unison. She then built a prototype to verify that her final solution met the original specifications.

Throughout the project Elizabeth kept meticulous records in her project journal. She ordered her parts as early in the project as possible. More than once she had to remind part suppliers of their promises, which she recorded in her project journal. She wrote monthly progress reports, gave oral presentations, and wrote a final project report.

The recorder project, like most engineering projects, progressed through the following series of stages:

1. Develop functional specifications.
2. Develop the concept design.
3. Generate design alternatives.
4. Select and model the best alternative.
5. Test and verify the design.

The success of every engineering project depends on keeping the client requirements or customer needs constantly in mind. All involved in the project—clients, customers, engineers, managers, marketing, and manufacturing—should provide input into the development of the specifications. As much concurrency as possible should be built into the project to speed its completion. As appropriate, engineers should use computers to aid in the design and simulation process. Engineers should investigate as many alternatives to the concept design as time and budget constraints allow. Designs for testing and manufacturing techniques should also be considered when investigating alternatives. Throughout every project, engineers should maintain complete and clear documentation.

Problems

1. Many nonengineering activities use steps similar to the five-step design process given in this chapter. An example of this is the decision-making process you use in making a major purchase, such as

buying a car. Discuss how the design process described in this chapter can be applied to the purchase of a car, identifying as many of the five engineering design steps as you can.

2. Interview a design engineer. Ask the engineer to describe the design stages he or she uses in his or her own design process. Compare those stages to the ones presented in this chapter.

3. Consider a hypothetical multidisciplinary business team consisting of representatives from engineering, research and development, manufacturing, marketing, finance, and purchasing. The team is working to develop a new bicycle that can be easily folded and stored in a small space, such as a car trunk or a locker. Suggest tasks for each team member that, together, would ensure the success of the project.

4. In the example given in problem 3, in which roles would you expect to find engineers? How would an engineering background contribute to their success in these roles? (Hint: As in Problem 1, focus on the use of the five-step design process in a nonengineering application.)

5. Investigate how each of the following items was developed, and speculate about what occurred at each of the five engineering design stages: space shuttle tiles, "Post-It" notes, nonstick cookware, antilock brake systems on automobiles, automobile air bags, and the Eiffel Tower. Your research sources can include encyclopedias, books, science magazines, technical journals, and/or resources available on the Internet.

6. Develop a list of functional specifications for your ideal house. Then draw a concept design for a house that fully meets your specifications.

7. Draw sketches of three alternative designs for the layout of your bedroom. Indicate the design you prefer. Identify a computer program you could use to model the layout of your room.

8. When possible, decisions should be tested and verified before they are implemented. For example, you might decide to buy a certain car, but you would probably test drive it before actually purchasing it. How would you test and verify a decision to attend a given engineering school?

9. List the documentation that accompanies a car, a VCR, and an engineering calculator. What kinds of information does each document give the purchaser?

10. What have your job experiences told you about yourself? What did you enjoy or dislike about your past jobs? For example, did you enjoy or dislike the responsibility of a baby-sitting job? The solitude of an early morning paper route? The fast-paced personal interaction of a job at a fast-food restaurant? The hard work of a furniture moving job? The challenge of an engineering internship?

2 The Engineering Disciplines

"I chose engineering because I liked to design cars," says Ashland Brown, Dean of Engineering at University of the Pacific in Stockton, California. "While I was in high school, I used to enter Fischer Body Craftsman Guild contests. These were the little model cars, 1/16 scale. I built six of them and won two college scholarships for my efforts." Three years after earning a Ph.D. in Mechanical Engineering in 1974, Brown began a 12-year career with Ford and General Motors before moving into academia. One of his favorite projects was managing the development of a road simulator for GM. The goal was to replace the costly, time-consuming method of testing vehicles widely used at the time. Brown explains: "At some point in the design of new cars, you have to devise a way not only to test the concept but also to prove that the new product will do what you anticipated it would do. Most techniques used at the 'Big Three' auto makers involved the use of proving

grounds. Proving grounds are huge investments of people, resources, and grounds that cover thousands of acres. They have hills, valleys, bumps, and all kinds of obstacles used to test vehicle systems and components. Testers drive a vehicle over and over and around and around the proving grounds, 24 hours a day, seven days a week.

"We worked with a company called Minnesota Testing Systems to develop a road simulator, a machine that went into a huge laboratory. You can put a vehicle on this machine, shake it down, calibrate it and run a simulated test in six weeks that would normally take about eight to nine months to do on the proving grounds, at a very small fraction of the cost. And if a vehicle breaks down, you can see right then and there what happened. By contrast, if a car breaks down on the proving grounds, the problem might not be discovered for another four weeks, when they do the tear-down on the vehicle."

The road simulator was an early step in the gradual computerization of the industry, Brown notes. "The road simulator enabled us to do three-dimensional analysis and testing and actually verify the vehicle's design by computer. What we were looking at was the future, when we could design entire vehicles and do the fatigue analysis on the computer." Today, every major automotive manufacturer uses computer-aided engineering (CAE) systems to design and test vehicles.

I asked Ashland Brown how being an African-American affected his early career in the automotive industry. He replied: "I'm sure it affected me in some way or other, but the way I saw things, I happened to be a bright engineer in a very white-oriented business. They didn't promote me because of my skin color or my ethnicity but because of the qualifications that I brought to the party—my problem-solving skills and ability to manage people. My boss and the chief engineer would not have handed me the crucial simulator project and others like it if they didn't think I was the guy who could get the job done."

INTRODUCTION

Ashland Brown's story illustrates just one of countless paths that a mechanical engineer's career can take. This chapter explores the wide variety of engineering disciplines available for study and also focuses on several stories of actual engineering projects in various disciplines told by the engineers who worked on them.

Nearly all engineering schools offer degrees in the three oldest and broadest disciplines—electrical, mechanical, and civil engineering—and the majority offer degree programs in chemical, computer, and industrial engineering as well. In addition, there is a broad range of specialized disciplines from which to choose, from aeronautical to ceramics to ocean engineering. While some students choose to earn specialized degrees, many more earn broader undergraduate degrees and specialize later. Figure 2.1 indicates the distribution of the more popular disciplines described in the chapter.

In this chapter we will review the major engineering disciplines and more than a dozen of the most popular engineering specialties. As you read about the many career opportunities that exist for engineers within each discipline, keep in mind that many engineering graduates accept positions outside their major fields. For example, electrical engineers may take positions in biomedical engineering, mechanical engineers may take jobs as petroleum engineers, and civil engineers may go into industrial engineering. Many engineers go into other fields, such as law, medicine, and business, for which an engineering degree offers excellent preparation.

Figure 2-1
Distribution of the More Popular Engineering Disciplines

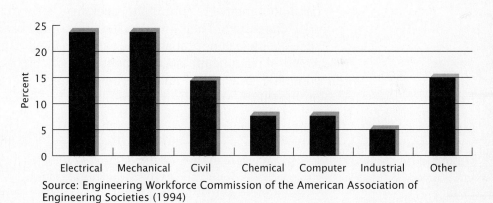

Source: Engineering Workforce Commission of the American Association of Engineering Societies (1994)

2-1 ELECTRICAL ENGINEERING

Electrical engineers design and develop such diverse products and systems as electrical power generation and distribution systems, radio and television broadcast and receiving equipment, home and recording studio audio equipment, worldwide telephone and related communications systems, microelectronics and microprocessors, and computers and their networks. Another popular, relatively new field is control systems engineering. Control systems are analog or digital electronic devices that control the operations of a broad range of products and systems, from robots to pacemakers to manufacturing processes.

Interest in electrical engineering exploded during and after World War II as electrical engineers developed inventions such as radar and contributed to the moon landing missions. Electrical engineers are now being challenged to develop a new worldwide communications system that uses optical fiber cables, microwave stations, and satellites to carry telephone, computer, facsimile (fax), television, and electronic mail (e-mail) to virtually every point on the planet. In urban areas, optical fiber cable will carry this information directly in digital form to every household, offering not only standard cable television but also e-mail, videotelephone service, and movies on demand.

Electronics-based industries, such as the radio, television, computer, automobile, and aerospace industries, employ enough electrical engineers to make them the largest engineering group. Such industries employ electrical engineers not only as designers but also as product and system testers, researchers, and sales engineers. Employment opportunities also exist with manufacturing and processing plants, industrial operations such as lumber enterprises, oil companies, railroads, and public utility companies.

2-2   COMPUTER ENGINEERING

Before computer engineering was established as a separate field, the computer engineer was often an electrical engineer or computer scientist who specialized, in the design of computer hardware and software. Today's computer engineers continually seek new ways to build faster, more powerful computers. Their goals include reducing costs, increasing processing speeds, increasing capabilities, and making computers easier to use. Areas of significant development include high-speed microprocessors, computer networking and internetworking, reduced instruction set computing (RISC), and optical computing.

FOCUS ON

ELECTRICAL ENGINEERING

"About a year ago I worked on a very exciting project for developing a system to detect and track bullets in real time. I designed most of the algorithms, did all the initial software simulations, designed the real-time data processing system, purchased all the equipment, went through an intensive training course to learn how to program the real-time hardware, wrote most of the real-time image processing software, and supervised two other software engineers to help with the software development.

It took about six months from the time I started working on the project to the time we were demonstrating our first prototype. The project was a complete success. Many government agencies are currently interested in funding this work further.

We think a good application would be in surveillance systems for high-security places like the White House. Or the system could be used in a war zone such as Bosnia. The system would identify, track, and compute the source of a bullet before it struck its target. This would allow defense personnel to attack the source of a sniper, hopefully before he can do much damage."

Jose E. Hernandez
Electrical Engineer
Lawrence Livermore
National Laboratory
B.S.E.E., University of
Puerto Rico

Computer engineers can specialize in many possible fields that involve the design of supercomputers, minicomputers, personal computers, local area networks, and wide area networks. A new and growing field is embedded system design, which involves the design of microprocessor-based systems that monitor and control products and systems such as VCRs; microwave ovens; and automotive ignition, fuel, and emissions control systems. Other opportunities include the design of computer peripheral equipment, such as computer terminals, printers, and network switching devices. Special challenges lie ahead for computer engineers as they continue to advance computer technology into the 21st century, meeting the information and processing needs of an increasingly technological world.

Computer engineers can find employment with government agencies and a wide range of companies, including not only computer manufacturers but also consulting firms and companies that create products that contain embedded systems.

As computer hardware costs continue to decrease, software has become the major cost component of many computer systems. New programming techniques, such as object-oriented programming (OOP), are being developed to attempt to offset this imbalance. OOP seeks to break large software projects up into more manageable modules (objects). Other interesting software developments include artificial intelligence systems that emulate different aspects of human intelligence. These trends have resulted in the development of a specialized field of engineering called software engineering. Software engineers oversee the design and development of large software projects. The title software engineer, however, is a controversial one. Some states forbid its use because no official engineering organization recognizes it.

2-3 MECHANICAL ENGINEERING

Mechanical engineers apply the principles of mechanics and heat to the design of machines such as internal combustion and jet engines, and thermal fluid systems such as those used in power plants. Their creations are

---

FOCUS ON

CHOOSING A DISCIPLINE

"I was always directed toward science and mathematics as a kid in school, and my father is in the nuclear industry. I grew up watching what he did, and I thought that would be an acceptable profession.

Initially I was heading into chemical engineering, but I found out it wasn't what I wanted to do. So I changed my major to mechanical and found that was much more interesting to me."

Clay Hess
Lead Engineer, Mechanical Hydraulic Systems Organization
Boeing Corporation
B.S.M.E., University of Washington

"I enjoyed chemistry in high school. I found it interesting and thought 'Maybe I can make a career of this.' I wasn't interested so much in the science aspect as in the application of chemistry. So I decided on chemical engineering."

Ian Buist
Vice-President
S. L. Ross Environmental Research Ltd.
B.S.Ch.E., University of Waterloo (Ontario)
M.S.Ch.E., University of Toronto

dynamic, involving motion, in contrast to the static creations of most other engineers. Mechanical engineering requires knowledge in a broad range of fields, including machinery, mechanics, thermal science, and instrumentation. Since modern mechanical systems are commonly designed using CAD/CAM systems, today's mechanical engineer must be adept at using computers.

Mechanical engineers are responsible for the design and development of an incredible range of systems and products. They design engines and power plants that convert chemical, nuclear, and thermal energy into mechanical energy. They also design energy transport devices such as gears, heat exchangers, and mechanical linkages. Still other products designed by mechanical engineers include household appliances, robots, mechanical tools, and transportation vehicles of all types. Mechanical engineers must be able to select the appropriate materials for the products they design—everything from vacuum cleaners to space shuttles—and ensure their structural integrity.

Mechanical engineering is considered by many to be the most general branch of engineering, providing a wide choice of careers and allowing movement into a variety of engineering and nonengineering areas. Mechanical engineers find employment in a broad range of industries, including automotive, construction, heating and cooling, manufacturing, power generation, space, and transportation.

2-4 CIVIL ENGINEERING

Civil engineers design and oversee the construction and utilization of major structures and facilities, such as buildings, airports, bridges, dams, flood control works, hydroelectric installations, aqueducts, pipelines, factories, and sewage treatment plants. They also help plan and construct roads, streets, housing subdivisions, highways, and railroads. A major problem area for civil engineers is how their projects impact the environment and the public as they seek technically and economically feasible ways to meet society's needs.

Many civil engineers specialize in structural engineering, which involves the design or construction of large structures such as buildings and bridges. The structural engineer may cooperate with an architect. While the architect concentrates on the aesthetic and functional aspects of design, the engineer is concerned with the materials and methods of construction. Another popular specialization is construction engineering, in which civil engineers manage the construction of other engineers' designs, concerning themselves with the scheduling and coordination of phases of construction and inspection to ensure adherence to specifications. Many civil engineers specialize in environmental engineering.

Civil engineers often find technical, administrative, or commercial positions with construction, power, transportation, and manufacturing companies. Government agencies at federal, state, and municipal levels employ large numbers of civil engineers to design and manage the construction of the nation's highways, roads, public transportation systems, and government buildings. Compared to the other engineering fields, civil engineering offers extensive international employment and consulting opportunities.

2-5 CHEMICAL ENGINEERING

Chemical engineering is a wide-ranging discipline that requires a broad education in areas such as chemistry, physics, materials, and electrical, mechanical, and industrial engineering. Emphasis is placed on the study of atomic, molecular, and crystalline structures, chemical reactions, and the physical properties of solids. A good example of a chemical engineering application is the development of less expensive, more efficient desalination plants. Another is the refinement of petroleum into its many derivative products, including gasoline, diesel fuel, kerosine, and asphalt.

Chemical engineers design and direct the production of fuels, plastics, fibers, paper, food, building materials, industrial chemicals, and pharmaceuticals. They are also involved in the design, construction, and operation of production facilities. Many of the chemicals they must deal with are toxic, so chemical engineers are often presented with special problems of pollution control and environmentally safe disposal of chemical waste products. Newer areas for chemical engineers include biotechnology, composite materials, environmental engineering, semiconductors, and superconductors.

Chemical engineers find positions in many different areas. Potential industrial roles for the chemical engineer include designer, supervisor, construction engineer, environmental engineer, and industrial engineer. The world's largest chemical producers, which are headquartered in the United States, employ large numbers of chemical engineers. Oil companies hire chemical engineers to design oil refineries and to develop techniques for cleaning up oil spills.

2-6 INDUSTRIAL ENGINEERING

Industrial engineering is primarily concerned with the analysis and design of more efficient production methods, including such tasks as selecting the necessary tools and materials and then designing a sequence of production operations. Industrial engineers also design plant facilities, establish work standards through time and motion studies, set up assembly and production lines, develop wage scales based on an analysis of required job skill levels, and determine quality-control procedures.

Major problem areas for industrial engineers are cost containment, productivity, and quality control. Other areas of concern include labor supply, machine utilization, provision of utilities and services, transportation of raw materials, and shipping of finished products.

An important specialty of industrial engineering is the study of human factors, or ergonomics, which focuses on the design of machines, processes, and environments that accommodate the physical capacities and limitations of the human body. Examples include the ergonomic design of automobiles and tractors, computer terminals and keyboards, commercial kitchens, and production assembly lines. To design applications such as these, industrial engineers must consider monotony, fatigue, information overload, and injuries such as carpal tunnel syndrome.

2-7 OTHER DISCIPLINES

The field of engineering is incredibly diverse. From the older disciplines of electrical, mechanical, and civil engineering, a long and growing list of engineering specializations have evolved. Some of these specializations can be considered branches of the three general disciplines, but others are not so easily classified. This section describes some of the more popular areas of engineering specialization, which combined form about 15 percent of all engineering professions. Except for aerospace, which represents about 4 percent, each field represents roughly 1 percent of the total engineering field.

Aerospace

Aerospace engineering is concerned with flight within and outside the earth's atmosphere. From its beginnings with the Wright brothers' primitive wind tunnel, aerospace has depended on scientific research for the knowledge base needed to design and build airborne and space-going machinery. During World War II aerospace engineering came into its own with the development of such relevant technologies as jet-powered aircraft, radar, and rocketry.

One branch of aerospace is aeronautical engineering, which is concerned with designing, building, and flying aircraft. Industries that have spun off from the airline industry and also hire aerospace or aeronautical engineers are involved in flight testing, flight simulation, robotics, controls, instrumentation, and exotic materials such as lightweight metal alloys and fireproof plastics and fabrics. Aerospace engineers can specialize in many areas, including aerodynamics, aviation electronics, hydraulic systems, navigational systems, propulsion, structural design, stress analysis, and even adhesives.

FOCUS ON

CIVIL ENGINEERING

"One interesting project we did was a geotechnical investigation and foundation evaluation on a public shopping center, after another engineering firm had completed the work. That firm concluded that a pile foundation was the only viable option.

We performed more soil test borings to define the soil strata, which was especially soft clay strata. We took undisturbed soil samples and conducted laboratory consolidation tests. (The other company did not perform these tests.) We worked with architects and structural engineers to better define the loading conditions. Then we performed settlement calculations based on soil boring, laboratory testing data, and loading conditions.

We found that the previous firm had over-engineered the foundation. In the end we recommended to the owner that he change the requirements of the shallow foundation bearing capacity from 2000 pounds per square foot to 1000 pounds. The architect and structural engineers reworked their designs, and shallow foundations were used satisfactorily, with savings in the neighborhood of $500,000."

Guoming Lin
Geotechnical Engineer
S&ME, Inc.
B.S.C.E., M.S.C.E., Hohai University, Nanjing, China
Ph.D., University of Tennessee

The complexity of modern jet aircraft and space vehicles requires engineers of all types to cooperate in the aerospace design process. For example, aerospace engineers often work with electrical engineers because both electronics and computer power are necessary to put a craft in the air or in space and bring it safely back to earth. Others involved in the aerospace industry include mechanical, materials, structural, computer, and systems engineers.

Since space travel is funded and controlled by the federal government, most aerospace engineers are employed either by the federal government or the aerospace companies it hires. Most aeronautical engineers are employed in the airline industry.

Agricultural

Agricultural engineers are concerned with the efficiency and economy of food production, mechanization, irrigation, labor availability and utilization, and the best use of available lands. A major area of emphasis is the production of a wide variety of safe, low-cost, desirable consumer food products. Other focus areas include soil chemistry, plant genetics, and the environmentally safe use of fertilizers and pesticides. Today many agricultural engineering degree programs have expanded to include biological and biosystems engineering.

Agricultural engineers not only developed the efficient, easy-to-use tractors and field equipment used in modern farming but also implemented satellite-based global positioning systems, which can relay the position of a tractor in a field to within a matter of feet, making precision farming possible. Agricultural engineers are also involved in the design of efficient concentrated livestock facilities that minimally impact the environment.

Many agricultural engineers move directly into farming, and others work as consultants to farmers. The huge American demand for inexpensive, convenient food products has created increasing opportunities for agricultural engineers in the food processing industry. Other career opportunities are found in the dairy, beef, pork, and poultry industries.

FOCUS ON

CHOOSING A DISCIPLINE

"I would say I was chosen to be in the engineering profession rather than I chose it. There was not much choice in terms of the field of study in China when I went to college. In fact, I knew little about engineering at the time when I graduated from high school. It is very fortunate for people to have a choice. However, I believe that people can succeed in virtually any profession as long as they are committed and work hard."

Guoming Lin
Geotechnical Engineer
S&ME, Inc.
B.S.C.E., M.S.C.E., Hohai University, Nanjing, China
Ph.D., University of Tennessee

"When I was a senior in college, I went to a conference with one of my professors. The conference was given by the state pollution control department. There was someone lecturing about hazardous waste, and I thought 'This is exactly what I want to do!' It's like a light bulb went off in my head. And I had never even taken a class in the subject before."

Rebecca O'Dell
Environmental Engineer
U.S. General Services Administration
B.S. Ch.E., University of Arkansas

Biomedical

Biomedical engineering is an interdisciplinary field that applies science, medicine, and engineering to the development of instruments, machines, and methods for studying and treating sick or injured humans and animals. Biomedical engineering as a distinct profession is relatively new; yet, due to its popularity, several universities offer degrees in the field. Still, many biomedical engineers receive the major part of their education in a related field such as medicine or electrical engineering.

A biomedical engineer usually has a graduate degree and specializes in one of three main areas: bioengineering, medical engineering, or clinical engineering. The bioengineer studies biological phenomena and identifies potential applications in areas such as agricultural and environmental engineering. Medical engineers design and develop medical instrumentation systems, such as CAT and MRI scanners, and artificial biological systems, such as limbs and hearts. Clinical engineers seek ways to improve the efficiency of health delivery systems. A special challenge for today's biomedical engineers is to find ways to help contain medical costs. Biomedical engineers are often employed by medical equipment manufacturers, hospitals, and research organizations.

Ceramics

Ceramics include a wide range of nonorganic, nonmetallic materials whose manufacture requires heating at high temperatures. Developments in ceramics have produced new types of materials that surpass alternative materials in strength, hardness, light weight, and heat and corrosion resistance. Important applications for ceramics are many, including brick and tile, glass, cement, nuclear fuel processing, lasers, and superconductors. Ceramics are also used in electronic devices such as capacitors, filters, transformers, microphones, phonograph pickups, strain gauges, and ultrasonic devices.

Ceramics engineers design ceramic materials and products, and develop and direct their production. Examples include superconducting magnets, lightweight airplane bodies, and superhard cutting tools and bearings. The diverse applications for ceramic materials creates a wide range of opportunities for ceramics engineers. Employment opportunities are varied, including the aerospace, automotive, chemical, electronic, and nuclear industries.

Environmental

Environmental engineers strive to minimize the effects that engineering activities have on the environment. The main tasks of environmental engineering are control of air and water pollution, industrial hygiene, noise and vibration control, and solid waste and hazardous waste management. The discharge of particulates and gases from vehicles into the atmosphere, the release of untreated industrial and household wastes into waterways, and our society's "throw-away" attitude toward solid wastes have resulted in pollution becoming one of the 20th century's major concerns, creating a strong demand for environmental engineers.

Environmental laws have been passed in an attempt to prevent or limit the adverse effects of pollution and environmental degradation. Technology has begun to help us solve some pollution problems. Automobiles and most factories now produce a fraction of the pollution they produced a few decades ago. However, the environment is as complex as it is inter-related. The environmental engineer must consider many facets of pollution and deal with such issues as land, air, and water pollution, as well as thermal, noise, and radiation pollution. The opportunities are numerous for environmental engineers as they design ways to solve the many environmental problems of society.

Many environmental engineers come from the civil or chemical engineering disciplines. Some obtain graduate degrees in environmental engineering. They often work for government agencies or for companies that have a potential for harming the environment. For example, oil companies employ environmental engineers to ensure that drilling sites harm the land as little as possible and that all applicable environmental laws are obeyed. Many environmental engineers are employed by environmental consulting firms.

Manufacturing

Manufacturing is the process by which raw materials or components are fabricated and assembled into finished products. Manufacturing engineering is related to industrial engineering but focuses on the actual manufac-

---

FOCUS ON

CHEMICAL ENGINEERING

"One of my areas of specialty is the burning of oil in place in open waters; this is referred to as in situ burning. We started using this procedure in the north, where it's the only option for cleaning up oil spills, because of the ice and cold and inaccessibility.

One of the key problems with ocean-based oil spills is that, under the influence of wave action and currents, the oil begins to take in water—it emulsifies. And for some oils, it's a very stable emulsion. The water content in the oil can reach as high as 80 percent. This presents tremendous problems. It increases the viscosity of the fluid by a factor of thousands. As the water content of a stable emulsion reaches 25 to 30 percent by volume, you can't ignite it.

So for the past four years, I've been working with a group of people from Norway and Alaska in a cooperative effort of circumpolar nations to apply emulsion breakers—standard industrial surfactant—to make these small, stable water droplets coalesce and fall out of the oil slick, reducing the water content. This would allow us to ignite the slick and burn it off. We have moved the technology from its beginnings to one that is starting to gain wider acceptance.

There are trade-offs involved. There's no question that in situ burning creates a huge, black column of smoke. But there's been quite a bit of research done on what's in that smoke, and what are the relative masses of the smoke versus the oil. And the research suggests that you are trading off a relatively small amount of air pollution for a very short period of time in exchange for eliminating the greater hazard of the oil."

Ian Buist
Vice-President
S. L. Ross Environmental Research Ltd.
B.S.Ch.E., University of Waterloo (Ontario)
M.S.Ch.E., University of Toronto

ture and manufacturability of products. Manufacturing engineers design manufacturing facilities, or design or redesign products to make them easier to manufacture. They also improve manufacturing productivity by developing improved products and methods of fabricating, assembling, and testing products.

Manufacturing engineers commonly use computer-aided design (CAD) to create new products and new factory systems. Using CAD and related advanced software tools, engineers can more easily design improved product parts, product assemblies, and the facilities that manufacture them. Computer-aided manufacturing (CAM) enables engineers to produce parts that are cheaper and easier to assemble, as well as facilities that are more efficient, safer, and require fewer personnel for their operation.

Manufacturing engineers find interesting careers in virtually all areas of industry as they seek ways to improve products and lower the cost of their manufacture. Some of the industries that hire manufacturing engineers are the automotive, chemical, food processing, and heavy equipment industries.

Materials

Materials engineers develop new materials characteristics that are significant improvements over existing materials. For example, they create materials or combine existing materials to create products with increased resistance to fracture, fatigue, corrosion, and damage. Materials technology comprises the selection, production, processing, and combination of raw materials in such a way as to produce alloys or composite materials that have the desired shape and specified properties for optimum performance, particularly in harsh environments.

A good example of a problem for a materials engineer is the spark plug. The spark plug has an inner conductor that carries electricity, a ceramic cover that insulates, and a machined-steel shell that holds the spark plug in the engine block. Each of these materials has its own electrical, thermal, and wear characteristics. This presents interesting challenges for the materials engineer, who must ensure that all parts of the spark plug deteriorate at roughly the same rate and that the expansion or contraction of one part doesn't cause the spark plug to malfunction.

Materials engineers work with metals such as steel and titanium, synthetics such as rubber and plastics, glass, ceramics, semiconductors, and a variety of other materials. They must be familiar with many fields of technology, including chemistry, physics, metallurgy, ceramics, and computers. They often work closely with engineers from other disciplines who require the development of materials that are adapted to special applications or unusual environments. Materials engineers find employment in many areas, including the aerospace, chemical, electronic, and petroleum industries.

Mining

Mining engineers deal with the discovery, extraction, refining, and distribution of minerals. Their main task is to get as much ore out of the ground as possible, as economically as possible. Mining operations exist in all 50 states, with Florida leading in the total amount of mineral mined, followed by Arizona, Minnesota, and Texas.

Mine safety issues are a source of contention in the United States and throughout the world, since mining accidents can result in a significant loss of life. The toxic wastes produced by mining and smelting operations also present major environmental problems. Both issues present special challenges for the mining engineer.

Nuclear

A little mentioned but still widely used alternative to obtaining power from fossil fuels is nuclear power. Nuclear engineers not only design and operate nuclear power plants but also are involved in the production, reprocessing, handling, and disposal of nuclear fuel.

Over 400 nuclear power plants operate worldwide in 30 countries, generating about 17 percent of the world's electricity. More than 100 operating nuclear power plants make the United States nuclear energy program

FOCUS ON

INDUSTRIAL ENGINEERING

"We built a $50 million manufacturing and distribution facility in Nevada. It became operational in September 1994. As the project manager, I was responsible for the technical development of the facility. I did the master planning and put the budget together. When the board approved it, I managed four internal engineers and an external consulting firm, the construction group, and the site engineering team. I managed the whole startup of the operation.

One of the things we were charged with in this plant was not to do things the traditional way. There was a lot of pressure to be successful without spending a lot of money.

One of the things we didn't do in a traditional way was the design of the waste water system. We put in a full-blown anaerobic waste water treatment facility to handle our waste water before we discharged it to the town.

We designed and built a system in which we would segregate the waste upstream, out of the process, capture it, and contain it. We studied the process to determine where we would most likely have material waste. We designed the drainage system in that area to go to a segregation sump, which was nothing more than a brix meter.

The more concentrated the waste drain, the higher the sugar content. So that brix meter would continuously monitor the stream, and if it went over a certain set-point, we would divert the waste to a storage vessel.

We had three or four of these devices strate-gically located throughout the plant. We captured the waste continuously, and every couple of days we'd have a tanker haul it away to a facility where it would be remanufactured into ethanol. Because we reduced the organic load so much through that process, we could treat the rest of the waste for pH and send it to the town.

The entire investment in that operation was $500,000 at most, and operating costs are virtually zero. We just pay the trucking fee; we don't pay for any disposal. Compare this to the traditional plant, which would have probably cost $5 million to build and hundreds of thousands of dollars to operate each year."

Greg Raco
Project Manager
Ocean Spray, Inc.
B.S.Ch.E., Northeastern University

the largest in the world, providing 20 percent of U.S. power generation. Nuclear power is now the second largest source of U.S. electricity, exceeded only by coal, which provides about 55 percent of the country's electricity. The nuclear fraction is growing and is expected to reach about 25 percent by the year 2000.

Increasing concern about how burning fossil fuels contributes to global warming has kept interest in nuclear power high. Nuclear power plants do not emit so-called greenhouse gases, such as carbon dioxide. However, the safe and permanent disposal of nuclear waste has become a volatile issue. Furthermore, incidents such as the meltdowns at Three Mile Island in 1979 and Chernobyl in 1986 have brought into question the safety of nuclear plants. The result of such concerns is an uncertain future for the nuclear power industry.

Still, the U.S. nuclear power industry generates significant employment opportunities for nuclear engineers. Nuclear engineers also find careers in the U.S. Navy, whose entire submarine fleet is nuclear powered.

Ocean

Ocean engineering is a cross-disciplinary field that deals with all aspects of the marine environment. Ocean engineers must have a broad knowledge of marine structures, vehicles, chemistry, geography, and ecology. In other words, their education combines civil, mechanical, chemical, and environmental engineering within a marine context. Clearly, the challenges are many as ocean engineers deal with problems such as ocean depths, weather at sea, and generally hazardous working conditions.

Ocean engineers collect, process, and interpret data that results from ocean studies, they design and construct large ships and ocean-based structures such as oil platforms. Since their work often involves research, they often have graduate degrees. Ocean engineers normally find employment with organizations that deal with the ocean, including government agencies, environmental organizations, and oil companies.

Petroleum

With the exception of food, petroleum may be the most important substance consumed by modern society. It provides the fuel for virtually all automobiles, trucks, trains, and airplanes, as well as fuel for the creation of electricity for heating and for industrial production. From petroleum come the raw materials for plastics and other synthetics, paints, roofing and other construction materials, fertilizers, insecticides, soaps, and asphalt.

Petroleum engineers deal with all phases of the petroleum industry, emphasizing reservoir discovery and oil and natural gas recovery, and also involving petroleum refinement and distribution. One problem petroleum engineers face is the need to develop methods by which the energy contained within remaining oil and natural gas deposits, oil shales, and coal can be economically converted into usable form. To do so, engineers will have to design new recovery and processing techniques. Another problem is meeting the demand for new and cleaner oil and coal that will have fewer harmful effects on the environment.

Since oil is found in every region of the world, most oil companies operate internationally, giving the petroleum engineer ample opportunity to work overseas.

Power

Power engineers develop systems that generate, distribute, and use electricity. They design equipment such as electric generators and motors, transformers, and high-voltage power distribution systems.

All current methods of generating power negatively impact the environment in some way, challenging power engineers to find alternative ways to generate electricity that result in less damage to the environment. For this reason many power engineers are involved in developing technologies such as hydroelectric, solar, wind, and geothermal power, as well as the more exotic alternatives of magnetohydrodynamics, electrogasdynamics, thermoelectrics, fuel cells, and tidal power.

The highly industrialized nature of the United States and the size of its population combine to make it the world's largest user of electric energy, creating a significant demand for power engineers in this country. Most find employment with public utility companies, including those that use nuclear power, and large industrial companies.

Systems

Systems engineers design and implement complex systems such as factories, interstate highways, and hospitals. One of the main problems they face is the coordination of the major components of such complex systems. Typical components include parts and materials suppliers, robots and other machines, and employees. Systems engineers also design the interfaces between the electrical, computer, mechanical, and other components in systems. For example, they would design the interconnection of the monitoring, instrumentation, and piping systems in a chemical production facility.

Once a system goes into operation, the systems engineer may continue to work with it throughout its useful life, providing suggestions on its modification and its expansion to meet changing needs. In that capacity the engineer's title may change to operations engineer. Systems engineers can come from almost any field of engineering, particularly industrial, civil, and mechanical. They find challenging careers with a wide range of employers, including manufacturing and processing facilities, complex systems such as governments, and organizations such as hospitals, which have a critical need to ensure that all departments and personnel work together effectively.

SUMMARY   This chapter described the tasks and opportunities of the more common types of engineers. and some of the more popular engineering specialties. Each engineering discipline can offer you an exciting and rewarding career as you design better, quicker, less expensive ways to use the forces and materials of nature to make people's lives healthier, safer, and more enjoyable.

Problems

1. This chapter describes a number of engineering disciplines. Choose the two disciplines you find most interesting, and write a one-page report on each. Include current information about the tasks performed by engineers in each discipline and information on job opportunities and income. To obtain this information, interview engineers or professors, search the library for journal articles, or search the Internet.

2. Choose two of the engineering disciplines described in this chapter for which you feel least suited. State your reasons for your choices.

3. The development of new technologies can stimulate new engineering disciplines. For example, computer and environmental engineering are relatively new disciplines. Name one or two new engineering disciplines that could possibly arise during your lifetime.

4. Select an engineering discipline, and write a two-page history of it. Identify events that influenced the current focus of this discipline. Your research sources can include encyclopedias, books, technical journals, and/or the Internet.

5. Interview one or more established, experienced engineers or members of your college engineering department. Ask them to describe how their engineering disciplines have changed since they received their bachelor's degrees, and report your findings.

6. Assume mechanical, electrical, and materials engineers are working as a team to design a new computer keyboard. For which parts of the keyboard would each engineer be responsible? What does each engineer need to know?

7. Write a three-page biography of a 20th century engineer who made a major contribution to a significant engineering development, such as the electron tube, radar, the jet engine, the ENIAC computer, the Golden Gate Bridge, the helicopter, or manned space travel.

8. Write a three-page biography of an engineer who founded a well-known company of the past or present, such as Westinghouse, De Lorean, Sperry Rand, or Underwriters' Laboratories.

9. Write a five-page description of a well-known engineering project, such as the Manhattan Project, the 1960s moon-landing project, the Sears Tower, or the Boeing 777 airliner. Describe how the project might have followed a five-step design process similar to the one described in Chapter 1.

10. Select an engineering discipline, and search the World Wide Web for the Web site of an engineering company that is associated with that discipline. What information does the company's Web page give you.

The Skills of the Engineer

"My work on the 777 posed some interesting engineering challenges," says Clay Hess. Hess managed a team that designed the installation of the hydraulic systems for the Boeing 777, the first aircraft designed entirely with computers. "Our job was to route all the hydraulic plumbing, if you will, throughout the entire airplane—roughly 1100 individual hydraulic hoses," says Hess. "We had to route the hoses down the main landing gear in such a way that it would not be damaged by foreign objects or material from the runway. We also had to make sure that the hoses would clear all the mechanisms that pull up as the gear comes down from the airplane. We had to go through quite a list of engineering questions to determine exactly what the best routing was."

After a year and a half, the team came up with a design they felt was adequate. The next step was to work with a group called Airplane Integration, whose job was to verify that every piece of the aircraft fit together properly. That's when a design problem was discovered, says Hess: "In the initial design, we had eliminated the kinematics because it takes so much computer effort and time. So we basically tested the design in the gear-down and gear-up positions. We verified all our clearances from those two positions. But when we started looking at the overall kinematic movement of the model, we found that, as the gear moved from one position to the other, the hoses would hit the structure. That required a significant redesign of some of the hoses on the aft part of the gear."

Clay Hess believes that teamwork and communication skills are critical to the success of any engineering project. "There's an exceptional need to be able to communicate and work well with the other individuals on a team," he says. "When you are in a team environment, you are talking to groups of people all the time. On the 777 project, presentations occurred at least once a week, and more often when the initial design was being done.

"I think that it certainly would have been to my advantage to have gotten that team experience in college, as well as the communication experience," Hess adds. "I spent a lot of time in the mechanical engineering curriculum, taking courses above and beyond my degree requirements. The one mistake I made was that I didn't spend enough time taking classes that built communication and teamwork skills."

INTRODUCTION   This chapter focuses on the skills you will need to succeed as an engineer. The skills are described in the context of a real-world project to develop a high-precision metal-cutting lathe and the systems (equipment and building) to support it.

As you read the chapter, you will discover that strong technical skills are not the only requirements for success. As Clay Hess points out on the previous page, engineers must also have excellent communication skills and the ability to work effectively as members of a team.

3-1   CASE STUDY: THE PVTM PROJECT AND TEAM

Most lathes, also called turning machines, operate horizontally. That is, the cutting tool moves in a horizontal plane while the part being cut spins on a horizontal axis (picture a table leg being carved on a woodcutting lathe). Horizontal lathes are typically used to cut relatively small, lightweight parts. Vertical lathes, on the other hand, are designed to cut large, disk-shaped parts on a rotating circular cutting table. The cutting tool moves in a vertical plane, up and down, from the center of the circular part to its edge. Such a lathe, sometimes called a vertical boring mill, is capable of holding and turning huge metal parts weighing many tons.

A vertical orientation is used because the parts that vertical lathes are designed to cut are so heavy that their shapes would distort significantly if they were hung on a horizontal lathe. A large, heavy part sags when mounted on a horizontal axis, becoming significantly thicker along its lower edge than along its upper edge. Under these conditions, the tolerances required of a high-precision lathe cannot be maintained.

The goal of the Precision Vertical Turning Machine (PVTM) project was to build a lathe that could cut parts to within a tolerance of one millionth of one inch (1 microinch, or roughly the thickness of 100 atoms in solid aluminum). This would represent an advance in vertical lathe technology of at least one order of magnitude beyond existing precision cutting capabilities. The lathe would cut metal parts of up to 64 inches in diameter, 20 inches in height, and 3000 pounds in weight. The PVTM is shown in Figure 3-1. A part being cut sits on the cutting table shown in the figure. The cutting tool, mounted on the tool bar, moves in the plane of the X- and Z-axes.

To achieve the desired tolerance capability, project engineers would have to solve four particularly difficult problems:

1. Accurately controlling the position of the cutting tool.
2. Accurately measuring the position of the cutting tool.
3. Minimizing seismic motion that could cause unpredictable movements in the lathe components and the part being cut.
4. Minimizing temperature variations that could cause changes in the size of the lathe components and the part being cut.

Figure 3-1
**The Precision Vertical
Turning Machine (PVTM)**

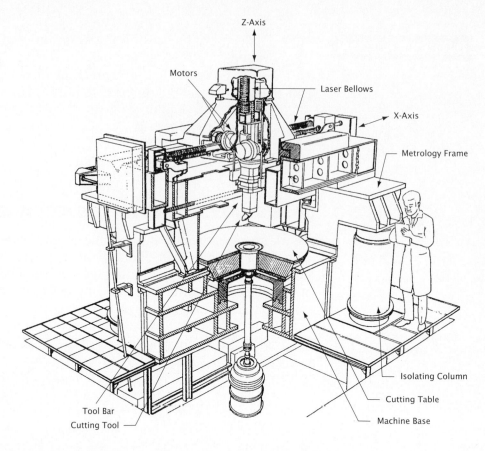

The PVTM project was conceived when an engineer at the company that developed it perceived an industry need. A large research grant from the federal government funded the project.

The company gave responsibility for managing the project to a project engineer with extensive experience in lathe design. He recruited three experienced engineers to lead the three subteams of mechanical, electrical, and civil engineers who would do the design work. The engineers were chosen primarily for their experience in designing lathe components and in computer modeling and simulation.

Including team leaders, the three subteams consisted of eight mechanical engineers (MEs); four electrical engineers (EEs), including one computer engineer; and four civil engineers (CEs). Also on the overall project team were three electronics technicians, four drafters, and four machinists. Including managers and secretaries, over 30 people were involved in the four-year, $15 million project.

The ME team was responsible for the design and development of the metrology frame, which acts as a reference for all cutting tool position measurements, the tool bar, the motors, the machine base, and the pneumatic isolators (see Figure 3-1). The team also worked on the design of the cooling system and the containment building, as described in Section 3-3.

The EE team was charged with designing and developing the cutting tool positioning systems, the cutting tool position measurement systems, and the computer interfaces and software.

The primary job of the CE team was to design and direct the construction of the containment building in which the lathe would operate. The CEs were also responsible for the installation of the lathe's support systems, such as the extensive air conditioning equipment and the coolant pumping system that would maintain a constant temperature environment for the lathe.

Once the team members had their assigned tasks, the project engineer met with the team leaders and worked out a budget and time schedule for the major parts (mechanical, electrical, and civil) of the project. The three team leaders then met with each engineer on their teams and worked out a budget and a time schedule for each of their assigned tasks.

3-2   MATHEMATICAL AND ANALYTICAL SKILLS

The PVTM project began with extensive mathematical analysis to determine exactly what would be required of the control and measurement systems. The ME team leader, with considerable experience in the design of vertical lathes and a strong background in mathematics, performed most of the initial analysis. Given the goal of 1 microinch tolerance and the size of the parts to be cut, she was able to determine the accuracy with which the cutting tool had to be controlled and its position measured. She also determined the maximum allowable expansion of the cut parts, establishing the extent of the measures that had to be taken to control part expansion, as described in Section 3-3.

The ME team leader also analyzed seismic data collected at the building site to determine the measures that would have to be taken to minimize lathe vibration. The team's zeal in eliminating external temperature and seismic effects was revealed during one of the early weekly team meetings when someone asked, "What about the moon? Will its gravitational effect distort the lathe or part so that it is pushed out of tolerance?" Good question. No one knew. The ME leader went to her calculator and books and came back within an hour. The answer, fortunately, was no.

Determining the electronic and computer control requirements was also a complex process. For example, the computer engineer determined, given the required accuracy of the cutting tool position measurements, that 64-bit electronics would be required of all position sensors and that 64-bit integer arithmetic would have to be used within the computer. Furthermore, 64-bit accuracy would be required when controlling the rotation of the motors that moved the tool bar on which the cutting tool was mounted.

Finally, simple mathematics showed that moving the cutting tool 1 microinch at a time over a typical 64-inch-diameter part would require 320,000,000 separate moves. The goal was to cut a part in three days, working around the clock. The computer engineer determined mathematically that moving the cutting tool 320,000,000 times in three days meant the PVTM computer had 1 millisecond to make each move. That is, once each millisecond, the computer must read the output of the cutting tool position sensors, calculate the position of the tool, read the next required position of the tool from a database containing a description of the desired shape of the part, and move the tool to the new position.

Engineers must always test the products, systems, and processes they design and analyze the test results to ensure that project specifications are met. As the PVTM engineers designed and constructed each component of the PVTM, they collected and analyzed test data to ensure that the components did the job they were intended to do. The position sensors were purchased as manufactured units but were tested for accuracy before being used. The ME team designed, built, and then tested the two motors that moved the tool bar along the X- and Z-axes, respectively. The EEs designed, built, and tested the electronics that collected the position information from the sensors and sent positioning signals to the motors.

As the engineers interfaced the PVTM components to each other, they tested each resulting subsystem and analyzed the test results to ensure correct interaction between the components. Of course, once the engineers completed the project, they tested the entire lathe and all its support structures, including the containment building. The results of these tests were analyzed to ensure that project specifications were fully met.

As an engineer, your most important tool will be mathematics. Your knowledge of math will enable you to apply scientific principles to the solution of engineering problems.

FOCUS ON

DESIGN

"Two of my most interesting clients are the largest and second largest carrot producers in the world. They have wonderfully automated carrot-handling equipment.

One problem my firm solved for the carrot producers involved the great quantities of water used for carrot processing. They wanted to put a waste water sump next to their building. The sump had to be concrete lined because the water must stay agitated so that dirt and heavy parts of the carrots can be separated and pumped out. It would be a small sump but deep.

Our original idea was to construct a reinforced concrete tank that would be 12 to 15 feet deep. But according to federal regulations, when you dig a hole that deep, you have to take special precautions to ensure that the sides of the hole do not cave in and injure the people digging the hole. The requirement also meant that we would have to tear down part of the building next to the tank.

We were weighing all the costs, and I said 'Why don't you put it in as a swimming pool?' So we dug the sump right next to the building. We did not have to tear

down any part of the building, and we saved the client over $30,000 in concrete costs alone. Because the sump has curved walls, rather than the 90-degree corners we had originally planned, the system was much more efficient and stayed cleaner because there were no corners to trap debris. The project was so successful that we have now built several more of the tanks in the same way."

Joe Engel
Structural Engineer
Engel & Company
Engineers

3-3 PROBLEM-SOLVING AND DECISION-MAKING SKILLS

In the course of any project, engineers must solve a series of problems and make many decisions about everything from where to source materials to which design option to pursue. The quality of those individual solutions and decisions will determine, in large part, whether or not the team achieves its objectives.

Problem Solving

As noted in Section 3-1, the PVTM engineers had to solve four key problems: controlling the position of the cutting tool, measuring its position, minimizing seismic motion, and minimizing temperature variations.

Controlling the Cutting Tool Position   To achieve a tolerance of 1 microinch, clearly the PVTM engineers had to be able to control the position of the cutting tool and measure its position to an accuracy of better than 1 microinch. To develop a method by which the position of the cutting tool could be controlled to this accuracy, the MEs started with a fairly standard metrology frame, tool bar, and cutting tool. The novelty for them was in the motors they designed to move the tool bar with the required accuracy. They designed two extremely low-speed motors that moved the tool bar in the X and Z directions, respectively. The motors were able to smoothly move the tool bar at a rate as low as 6.3 inches per year. At maximum speed the motors moved the cutting tool 1 milli-inch per second, or about 4 inches per hour. Again, the lathe computer used 64-bit electronics to control the positions of these motors.

Measuring the Cutting Tool Position   To meet the specification for measuring the cutting tool position, the MEs knew they would need very precise and accurate sensors. The initial concept design called for two laser interferometers, one each for measuring the X- and Z-axes positions of the tool bar. On the shaft that rotated the cutting table, they planned to mount a spindle encoder that measured the angular position of the table and the part sitting on it. However, some analysis of the data that these three devices would generate showed that it was not enough to allow the computer to determine the tool position with the specified accuracy.

The engineers generated several other potential design solutions and ultimately used 13 position-sensing devices to measure the position of the tool bar and therefore the cutting tool attached to it. Seven laser interferometers reflected laser beams at various places on the tool bar to determine the distance from the interferometers to the bar. The engineers mounted the interferometers on the metrology frame, which acted as a reference point. To increase the accuracy of the interferometers, the laser beams traveled in accordionlike evacuated laser bellows tubes (see Figure 3-1) that prevented air from slowing the beams of light.

The MEs mounted five differential capacitance gauges around the perimeter of the cutting table to measure any tilting of the table from the horizontal. As stated earlier, once each millisecond, the computer sampled the output of the 13 position sensors, calculated the tool position from their data, and moved the tool to its next position. Tests showed that the final design easily met the specifications.

Minimizing Seismic Motion   As previously stated, seismic measurements showed that vehicular traffic near the containment building could upset the lathe's accuracy. To isolate the vibration, the MEs designed and developed a pneumatic isolating column, 3 feet in diameter (see Figure 3-1). They mounted the lathe on four of these isolators. While apparently a simple solution, the result was a lathe that "floated" above its own foundation, contained in a building but not a part of it.

Minimizing Temperature Variation   Perhaps the most serious issue in accomplishing the basic PVTM project goal of a 1 microinch tolerance was metal expansion and contraction due to temperature changes. To minimize temperature variations, the temperature of the lathe's metal parts was held to within 0.1 degree Fahrenheit by building passageways throughout the lathe and then pumping temperature-controlled coolant through the structure. The effect was similar to that of a car's radiator. To minimize the effects of the minor temperature variations that still occurred, the mechanical engineers specified special metal alloys for the metrology frame and other metal components of the PVTM. These alloys had very low coefficients of expansion.

A $5 million, seven-layer building was constructed to house the lathe. Each of the seven building layers, from the outdoors to the lathe, used increasingly accurate air conditioning units to maintain temperature control. The air in the small room in which the lathe operated was controlled to within 0.02 degree Fahrenheit. The actual temperature in the room was not so critical; however, it was extremely critical that the temperature not

FOCUS ON COMMUNICATION

"The most interesting part of the road simulator project (see Chapter 2) was selling it to the vice-president. Scientists have a tendency to explain theories, but a vice-president wants to know the bottom line: "What is this really going to do for me?"

I practiced for at least a month for this $25 million presentation. My boss went over it, took all the scientific items out of my slides, and put in more dollars and cents. But what I think really sold the vice-president was when I told him the

Japanese, who were our major competitors, had 20 of these simulators. I explained that there are no proving grounds in Japan because they don't have the real estate."

Ashland Brown
Dean of Engineering
University of the Pacific

vary. The engineers designed extensive sound-proofing material into the building to prevent loud external sounds from reaching the lathe.

Three years into the project, while the MEs and EEs were testing their designs in a nearby building, the CEs had completed the PVTM containment building and were supervising the installation of the coolant pumping system that pumped water through the lathe and the extensive air conditioning equipment that cooled the seven layers of the building.

Decision Making

The three PVTM teams had to make several important decisions to achieve their goals, including decisions about materials and equipment. Some of the decisions involved trade-offs between cost and quality. For example, the MEs specified a metal for the metrology frame that was very expensive compared to alternative metals; however, the team concluded that the expense was necessary to meet the project specifications. Similarly, the EE team decided to purchase an expensive minicomputer after the computer engineer on the team determined it was the only machine that could meet the required 1-millisecond turn-around time to measure the cutting tool position and move it to a new position. The civil engineers spent hundreds of thousands of dollars on the required air conditioning systems.

Engineers must decide not only how much to spend to accomplish a goal but also from whom to purchase needed supplies, parts, and products. Engineers who attempt to save money by using a less expensive source may find themselves waiting for parts while a project falls behind schedule.

Often the major obstacle an engineer faces during the decision-making process is weighing the trade-offs among conflicting objectives. Can the project schedule slip a few days if it will save $5000? Is staying on schedule important enough to justify hiring an outside consultant at considerable expense? Which of the potential design solutions is best and should be adopted? Do these test results really indicate we have met our specifications? Who should design the interface between two system components—engineer A who is immediately available or engineer B who is better at such tasks but is already overloaded with work?

In addition to the many financial, technical, and personnel decisions engineers make are the ethical decisions. Should I cut the cost of this project as my client insists, even if it could result in an unsafe product? Should I overdesign this product to be absolutely sure it meets specifications, even if I must overcharge the customer to do so? Should I use confidential information that I picked up while working for a previous employer? These and other ethical issues are discussed in Chapter 5.

FOCUS ON

COMMUNICATION

"Our engineers generally have to give a written progress report every week, just a brief one-page summary of 'what we accomplished, what we're planning to do.' We require these reports partly because our clients generally want to see them. But it also helps us as managers to understand exactly where the engineers are on a project, what are their problems and concerns, and what they need from others."

Paul Olstad
Project Manager
Fluor Daniel

Nearly all engineers function as a part of a team. In that context many decisions are made as a group after extensive discussions. All decisions, whether made individually or collectively, must be consistent with project objectives and customer needs. This will make some decisions unpopular with one or more engineers, perhaps with you. Engineers, however, must always hold the objectives of the project above their own interests.

3-4 COMMUNICATION SKILLS

For an engineering project to succeed, team members must be able to communicate effectively with one another and with all those who have a stake in the project. The larger the project, the greater the need for excellent verbal and written communication skills. For this reason, many engineering schools have incorporated communications in certain engineering courses, and some schools require students to take separate communications courses.

Verbal Communication

During the course of a project, engineering team members are often required to give oral presentations, usually at project milestones. Engineers usually give their first oral presentation when they have chosen a concept design for the problem at hand. This initial presentation gives engineers the opportunity to present their ideas and possibly to gain new ideas from those who attend the presentation. Managers and other engineers usually attend this presentation to ask important technical and economic questions about the approach.

Having to defend their ideas and answer questions during oral presentations helps engineers to clarify and better understand their work. It may even cause them to abandon an approach or design before much time and money have been invested in it. The audience at the presentation may then suggest alternative approaches.

The Precision Vertical Turning Machine team worked closely together in a single building. All of the parts of the lathe and its support systems had to work together in such an intimate way that the communication among the engineers was nearly continuous. As a result, the team leaders did not require much in the way of formal communication.

The entire PVTM project team, however, did meet weekly to review the overall progress of the project and to discuss any problems. Each meeting generally lasted one to two hours, with the project engineer and team

FOCUS ON

DOCUMENTATION

"Many of our projects are what we call upgrades or rebuilds of existing facilities. So you often get a design that was created 2 years, 10 years, or 20 years earlier. You have to try to figure out what is there before you can decide how to make modifications to it. This is easiest to do when you have proper documentation.

It's essential to document what you've done on a project, and not just with formulas. Put in a few words now and then so that you can understand a year later what you were doing at the time."

Paul Olstad
Project Manager
Fluor Daniel

leaders leading the discussion and making notes and drawings on a white board. It was important that all team members attend these meetings. All were expected to listen and learn, and to voice their problems and concerns.

It was particularly important that the machinists and technicians attend these presentations. Since it was their job to create the engineers' designs, it was vitally important that they understood, and were able to implement, those designs.

When an engineer finished the design of a major component of the PVTM project, the team leader required that individual to give an oral defense of the design. Such oral presentations lasted from 20 to 60 minutes, depending on the number of questions asked by the team members who attended. The speaker used transparencies, often depicting machine part blueprints, system diagrams, or schematics, to illustrate a design. Sometimes the speaker used a workstation to present and/or simulate a computer model.

Written Communication

Engineers must be able to communicate their ideas clearly in writing so that other team members and those outside the group can understand and act on them. One type of document engineers are required to write is the weekly or monthly progress report.

All members of the PVTM project teams were required to write monthly progress reports. The reports let the team managers know what their design team accomplished during the past month so they could report that information to their managers. The progress reports kept everyone involved in the project up to date and helped keep the project on schedule and within budget.

Most managers prefer concise progress reports, usually one to two pages in length. Managers do not have the time to read lengthy progress reports. As an engineer, you must learn to say much with few words. Each manager is different, however, and you will have to match the style and size of your reports to that desired by your manager. An example progress report is shown in Figure 3-2.

Another required document for any engineering project is the engineer's final project report. This document details every aspect of the engineer's work on the project. The key purpose of the project report is to enable an engineer, in the future, to modify the product, process, or system that was developed during the project. Since this document is usually much longer than a weekly or monthly progress report, creating it is a time-consuming process that many engineers would prefer to avoid. The document is critically important, however.

Figure 3-2
**An Example
Progress Report**

To: Gary Henderson
From: Julie Ackerman
Subject: July PVTM Laser Interferometer System Design Progress Report

A. Introduction
Work on the laser interferometer system progressed well during July. However, I discovered that my originally specified interface chips are out of stock, forcing me to redesign the interface to accommodate available chips. Fortunately, testing has shown that the interferometers will fully meet our specifications.

B. Accomplishments
   1. Ordered new interface chips ("guaranteed" to be here by next week).
   2. Redesigned the interferometer system interface.
   3. Completely tested the interferometers.
   4. Documented the interferometer test results.
   5. Oversaw the technician's wiring of the interferometer control and data paths.

C. Work Remaining
   1. Construct the new interferometer system design.
   2. Test the interferometer system.
   3. Complete the interfacing of the interferometer system to the lathe computer.

D. Problems
   1. Time. I'm a bit behind schedule due to unavailability of some parts.
   2. Manpower. I could use another technician to assist in the wiring effort.

As indicated by the example table of contents in Figure 3-3, a project report usually begins with an abstract, which briefly describes the project goals; a table of contents; and a list of illustrations. The body of the report consists of a specifications list, a brief description of the concept design, and a detailed description of the design. Block diagrams, drawings, schematics, tables, and other figures are often included. The body of the report ends with a summary of the conclusions reached by the engineer.

A set of appendices should follow the body of the project report. Appendices usually include test data, in tabular and graphic form, that verify that the project meets its specifications. They also often contain blueprints, CAD drawings, sketches, schematics, program listings, graphs, parts and materials specifications, and instructions on how to use or modify the design.

The report should conclude with a list of reference materials, if appropriate.

Most engineering designs must be modified at some later date, and the project report is intended to ease this process. As you write the project report, consider its purpose. Two years from now, it may be you who has to modify the design it describes.

Figure 3-3

Table of Contents for a Data Recorder Project Report

Hardened Data Recorder
Table of Contents

At the end of the PVTM project, each engineer wrote a complete project report, detailing the specifics of his or her design work. These reports averaged about 50 pages in length. They explained, among many things, the reasons for the decisions made in the design and how to build, use, and modify the design. An important feature of these reports was the documentation of the extensive testing done to ensure each design met its specifications.

3-4 TEAM WORKING SKILLS

Perhaps the most important skill you will need to succeed as an engineer is the ability to function effectively as a member of a team. Being a team player requires solid interpersonal and interdisciplinary skills.

Working as a team member involves compromise. Engineers are often called upon to let go of what they consider great ideas. A manager may veto an idea, or the team as a group may decide an idea is not workable. Team members must be willing to acknowledge the experience and judgment of managers and fellow engineers. They must be committed to the team effort, even when, at times, their work load is increased or they feel a situation is unfair.

For example, the civil engineers on the PVTM project chose a location for the containment building that was near a road on which heavy trucks sometimes traveled. This created special problems for the MEs, who would have to design and implement pneumatic columns that would isolate the lathe from the seismic vibration caused by the trucks. However, they agreed to do so after the CEs convinced them that the location was ideal in every other way.

The MEs on the PVTM project wanted to use air bearings in the lathe because they would allow virtually drag-free movement of the tool bar. Eliminating friction would prevent the creation of heat that would cause metal parts to expand. However, the EEs wanted the MEs to use oil bearings, which have a small amount of drag. The drag, while producing heat, would make the system more stable, allowing more accurate tool position control and measurement. The MEs compromised and used oil bearings.

The computer engineer and one of the EEs had to compromise on the interface between the cutting tool position sensing devices and the computer that controlled the position of the cutting tool. The computer could read only 32-bit numbers via a parallel port, yet the electronics associated with the position sensing devices generated 64-bit numbers that reflected the position of the cutting tool. Because the EE seemed to be the busier of the two, the computer engineer designed the interface, which read and stored the 64-bit numbers and then passed them on to the computer in 32-bit blocks. She then wrote the necessary software for the computer to read the 32-bit numbers from the parallel port and reconstruct them as 64-bit numbers internally.

In addition to having the willingness to compromise, a team player must have a broad knowledge base that extends beyond an engineering major. Engineering subjects taken by students outside their majors often

---

FOCUS ON

TEAMWORK

"I believe there are two main challenges in having a group of people work together as a team. First, the work has to be broken down into small, manageable tasks so that you can delegate responsibility to the individual people working on the project. Second, there has to be a good technical match between the task and the person responsible for doing the task.

In the bullet-tracking project (see Chapter 2, FOCUS ON ELECTRICAL ENGINEERING) everything worked out very well since I knew how to break the complete system into smaller pieces, and I also knew the people and their technical backgrounds relatively well. This allowed me to predict fairly accurately how long it would take to complete the project pieces so that they all came together at the end."

Jose E. Hernandez
Electrical Engineer
Lawrence Livermore
National Laboratory

include electronics, instrumentation, robotics, materials, and computer-related subjects.

Having a wide array of interdisciplinary skills will bring you several benefits as an engineer. Most importantly, as you understand and take into account the requirements of other engineers, your work will more likely integrate well into the overall project. Interdisciplinary knowledge will also help you empathize with other types of engineers and their problems.

Having a good understanding of other disciplines will enable you to work on tasks outside your specialty, making you a more valuable team member. For example, the ME who designed the low-speed motors on the PVTM project had little experience in this area. He would have preferred to have the task given to someone else, but the team leader assigned it to him. He began by reading some books and journal articles about the topic. As he worked on the design, he occasionally asked advice from his team leader, who did have experience in this area. In another instance, when one of the EEs suddenly quit the project, the computer engineer had to study the EE's work, complete it, and write the required documentation. The EE had finished over 90 percent of his task but had kept poor records of his efforts, so the computer engineer had a difficult job on her hands, sorting out the work of the EE.

Not all engineers on the PVTM project were required to do design work outside of their disciplines. But they all had to understand the work done by team members in other disciplines so the designs of all three teams integrated well into the overall project. For example, the MEs had to understand the functioning of the air conditioning system the CEs built so they could design a cooling system that would function effectively within the system. The ME who designed the low-speed motors had to understand the work of the EE who designed the motor drivers, as well as the work of the computer engineer who wrote the software interfaces. For the project to come together in a seamless manner, each engineer had to understand how his or her work fit into the overall project scheme.

SUMMARY

At the end of the four-year PVTM project, the National Institute of Standards and Technology of Gaithersburg, Maryland, used the PVTM to cut a part and then attempted to find a flaw in the part. Their best equipment, the most accurate and precise in the world, could not measure any variation in the part from its specified shape. It was declared to be cut within a 1-microinch tolerance.

The PVTM project succeeded in large part because the engineers involved had a variety of skills that are crucial to the successful completion of any engineering project: strong mathematical and analytical skills; a well developed capability for solving problems and making decisions; excellent communication skills; and perhaps most important of all, the ability to work effectively as members of a team.

Problems

1. Describe the skills required of the PVTM team members during each of the five design stages. For each stage list the difficulties the team members could have encountered, including technical, financial, and interpersonal issues.

2. It has been said that if automotive engineers were keeping up with computer engineers, cars would cost $5 and get 1000 miles per gallon. What kinds of real-world constraints limit automotive engineers, but not computer engineers, so that this is not the case?

3. Do a brief study of the traffic control system in your city. Write a five-page report describing the deficiencies of the system and how it might be improved. Include ideas on traffic light design and placement and law enforcement. Discuss costs and other problems related to the implementation of your ideas.

4. Assume you were assigned the task of solving the one problem with automobiles that bothers you the most. What would the problem be, and how would you solve it?

5. Describe a time when you had to make a difficult decision. What made the process difficult? On what basis did you ultimately make the decision?

6. Think of your instructor as your boss and yourself as an employee hired to attend classes and take notes. Write a one-page progress report of your work for the last two weeks. Use Figure 3-2 as a model.

7. Write a project report about a problem you solved. Describe in detail exactly how you solved it, using Figure 3-3 and the five-step design process as guides. Keep the report short, two to three pages.

8. Interview a practicing engineer who works or has worked on teams. Ask about the challenges and rewards of teamwork. Write a one-page report about your findings.

9. Consider a major purchase you made that involved a financial commitment (a loan or gift) from someone else. What personal skills and objective evidence did you use to gain the aid of your benefactor?

10. Prepare a three-minute speech in which you introduce yourself to a group of strangers. Practice this introduction alone or in front of friends. Time yourself. Does your introduction convey the professional image that you want to present? Present yourself to your class.

4 Preparing for an Engineering Career

"The key thing that I look for when I interview students is an intuitive understanding of the material," says Tom Henderson, an on-campus interviewer for a Fortune 500 computer manufacturer. "I'm looking for somebody who has a good feel for engineering and for the process of making good trade-offs and decisions."

When asked how he determines whether students have a good grasp of engineering basics, Henderson stated, "I try to get them to talk about a class where they've either had big homework assignments or projects, or perhaps

about a co-op experience. I ask them to explain it, and then I ask 'Why did you do that?' or 'Why didn't you do some other things?' A lot comes out when people explain what they did on a project.

"The next thing I look for," Henderson says, "is someone who has good teamwork and leadership skills, someone who can work with people and can deal with conflict—that is, someone who can work on a team of engineers. And I look for a certain amount of 'get up and go' and enthusiasm. I especially look for people who are excited about engineering and their career choices."

How does Henderson decide which students to call for an interview? "I get a stack of resumes about three inches high," he says. "I take them home, and I pull 15 or 20 resumes that I'll call for an interview. As I scan through them, trying to make a first cut, I spend less than 30 seconds per resume looking for certain features: What's their GPA? Have they done any projects that are similar to what I'm hiring for? (That's good evidence that they're interested in doing the work I need done.) Do they have any classes that are specialized—graduate level classes, for example—or something else that makes them stand out?

"A lot of students think that they're broadening their options by not committing to anything—by taking one digital, one power, and one analog course, for example," says Henderson. "But in some ways they're making themselves more average, so they won't stick out of the pile. It's the person who perhaps has three digital courses and a networking course that would catch my eye.

"Holding an office in any kind of organization is a real plus," he adds. "Just because a student belongs to the IEEE may not mean much, because there are always people who sit in the back row and do nothing. It's actually doing something—being the float committee chairperson, anything that shows leadership—that's important."

The thought of interviewing for a job can provoke anxiety even among the best prepared students. But Tom Henderson wants students to know that interviewers are eager to see them succeed: "On-campus interviewers are hoping to see as many good students in one day as possible, so they are rooting for everyone to do well. Students shouldn't think 'This interviewer is trying to tear me down,' but rather 'This person really wants me to do well.'"

INTRODUCTION

T his chapter shows you how to prepare for your engineering career. It describes a typical engineering curriculum and emphasizes the importance of getting engineering-related work experience and being involved in engineering societies while in school. This chapter also offers advice on seeking your first job, whether it's a temporary or part-time job that you seek while in school or a full-time job you seek after graduation. The advice includes descriptions of effective resumes and cover letters, as well as good interviewing skills. Salary and employment trends and career opportunities are discussed, and the principles for planning an engineering career are presented.

4-1   GETTING STARTED

You are already well on your way to an engineering career just by being accepted to an engineering school. Your engineering school has probably given you a wealth of material to help you plan your education and ensure you meet all requirements for the degree you are pursuing. This section should help you to understand that material better. The section also describes opportunities for work experience that you should pursue, as well as engineering societies you should join, while in school.

College Education

At most engineering schools, all students take basically the same courses for the first two years. These courses are chosen to meet the recommendations of the Accreditation Board for Engineering and Technology (ABET), which accredits engineering degree programs in the United States. ABET requires that accredited engineering curricula include courses in mathematics, science, humanities, and social sciences, in addition to courses in the accredited major. ABET also requires engineering students to complete significant course work in those areas of study. The number of prescribed units totals about three years of instruction. ABET allows engineering schools the freedom of deciding how to allocate the remaining year or more of instruction that is required to receive a typical engineering degree.

The ABET recommendations for the major subjects are listed in Table 4-1.

One basic purpose of your education is to give you an understanding of the behavior and characteristics of the natural forces and materials you will be working with. To be successful as an engineer, for example, you must understand calculus as a description of nature. Merely learning cookbook procedures for solving standardized problems will not do. A study of science also helps you develop an intuitive feel for the world around you. Science teaches you how to examine nature and to make sense of its behavior through your knowledge of its underlying principles and structure. Your study of mathematics and science will lay the groundwork for your study of engineering, which is why math and science come first in your curriculum.

There are two phases to your study of engineering: one dealing with general engineering and the other dealing with your engineering major.

Table 4-1 ABET Engineering Education Recommendations

Subject	Required Courses	Recommended Courses
Math	5 or 6 semester courses, including differential and integral calculus, differential equations	Probability, statistics, linear algebra, numerical analysis, or advanced calculus
Science	4 or 5 semester courses, including chemistry and physics (at least a 2-semester sequence of study in either area)	Life sciences, earth science, advanced chemistry or physics
Engineering	Engineering science (courses that are rooted in mathematics and science but carry knowledge further toward creative application), and engineering design (courses that teach design as an iterative decision-making activity that results in a system, component, or process that meets a desired need)	Advanced engineering courses in one's major; general or introductory engineering courses outside one's major
Humanities and social studies	5 or 6 courses that provide depth and breadth, not just introductory topics (courses should reflect a rationale or fulfill an objective appropriate to the engineering profession)	Economics, business, English, sociology, political science, history, foreign languages

Your study of general engineering will include not only introductory courses, such as the one for which this book was written, but also courses in fields of engineering outside your major. Your understanding of the other fields and how they interact with your field will help you cooperate with other types of engineers as you work on large and complex projects as a professional engineer.

Most schools are specific about the courses you must take in mathematics, science, and engineering; however, they usually give you considerable leeway in your choices of humanities and social science courses. You should select courses that fit your goals. For example, business courses would be appropriate if you plan to go into consulting. If you expect to work internationally, you should take an appropriate foreign language. Other courses you should consider include computer science, economics, government, law, premedicine, and sociology. Planning ahead in this way is particularly important if you plan to pursue a career in a field other than engineering. Many people use an engineering degree as a starting point for other professional careers, such as business, law, and medicine.

Work Experience

All engineering schools encourage students to gain some relevant work experience before graduation. Most offer some assistance in the search for such experience, which may be in the form of a cooperative education job, a summer internship, or simply a part-time evening job.

Cooperative education employment involves the cooperation of engineering schools and private or public industry to provide relevant work experience for students. Co-op enhances an engineering degree program by relating theory to practice. While on co-op, students learn by doing, applying what they have learned in the classroom to a working situation.

A co-op job offers the student not only valuable experience but also an opportunity to check out a prospective employer. Likewise, the companies that hire co-op students get a chance to check out potential employees.

Most engineering schools offer cooperative education employment opportunities in some form. Typically, co-op jobs are optional, lasting for a summer, a semester, or even a full year. Some schools make a co-op a requirement for graduation and employ personnel to manage the program. Such schools often require an entire year of cooperative experience for graduation. The year may be divided into two periods, during which the student works for either one or two companies.

Not all engineering schools offer formal cooperative education opportunities, but most have some program that helps students gain work experience. Such experience is most often in the form of summer internships with local companies. Some schools strongly encourage seeking internships and have a list of companies that want to hire students.

Whether or not your school offers any help in your search for a job, you should seek engineering-related work experience during your years in engineering school. Such jobs will put your engineering studies in perspective, giving you an opportunity to see how engineering is done in the real world. Even if your summer or evening jobs are not relevant to engineering, at least they allow you to document in your resume your willingness to work.

A growing trend among students seeking engineering-related work experience is taking temporary jobs in foreign countries. The increasingly global nature of the engineering profession adds considerable value to an overseas work experience. Most overseas jobs do not require you to speak a foreign language. However, the ability to speak the language of the country in which you are working is an asset and may help you land your first full-time job overseas. Significant cooperative opportunities exist throughout Europe and Japan. Section 4-4 discusses international career opportunities.

If you are interested in a co-op job, internship, or other engineering-related work experience in the United States or in a foreign country, ask for information from your faculty advisor, professors, department chairperson, or the dean's office. Most engineering societies also offer summer internships; check with the societies' faculty advisors or student presidents. Nearly all universities have a career services office where you can find a list of part-time and summer jobs and can receive training in resume writing, interviewing, and job searching. Career services offices also provide information on international jobs. See your school catalog.

Student Engineering Societies

The student chapters of engineering societies have much to offer you, including the opportunity to interact with students with common interests. At student chapter meetings, members discuss important events and developments in their area of engineering. They often invite guest speakers, including practicing engineers who can help you relate your studies to the real world. Student chapters also organize field trips to local companies so you can see engineers at work. Nearly all engineering societies offer scholarships to their student members. Financial support for student chapters of engineering societies normally comes from the host university, the national office of the host society, and from fund-raising activities. Student members typically pay dues to the national office of the host society.

All student engineering societies have student officers. The tasks of student officers include arranging the times and venues of meetings, obtaining refreshments for meetings, finding guest speakers, organizing field trips, and conducting fund-raising activities. The president or chairperson leads the meetings and delegates tasks to the other officers. A secretary takes minutes at meetings, and a treasurer handles the group's finances. Most student chapters will welcome your willingness to be an officer, giving you an opportunity to develop leadership skills that will help you succeed in your career.

If you are an exceptional student, you are likely to be invited to join one or more engineering honor societies. Each engineering field has its own honor society. The premier honor society, which covers all fields of engineering, is Tau Beta Pi. Eligibility for membership depends on your class standing in your junior and senior years. Join if you are invited to do so. Honor societies offer many of the same benefits as other engineering societies, including interaction with other engineering students, guest speakers, field trips, and scholarships. Active involvement in honor societies—for example, as an officer—will give you important experience as a leader of leaders.

Be sure you attend the meetings of the societies you join and get involved. Consider it an important part of planning for your career, your job interviews, and writing your resume. You may feel a little uncomfortable about getting involved during your freshman year, but certainly you should not wait beyond your sophomore year. When employers read resumes, they typically want to see membership in one or more engineering societies. If they do, they are likely to ask you, during a job interview, about your experience as a member. For example, they might want to hear about what you learned from guest speakers and during field trips.

FOCUS ON

HOW I GOT MY FIRST JOB

"For my last co-op, I worked for a food manufacturer. Then, when I graduated, they were doing a very big expansion project, and I walked in to become the project manager. It was unheard of for someone coming out of school, 25 years old, to take on a $5 million project. But thanks to my co-op work there, they knew me; I'd done projects in that facility before. The relationships were there, and I understood the business. It was a good fit. That project lasted about a year and a half."

Greg Raco
Project Manager
Ocean Spray, Inc.

"While I was in my senior year at college, I got interested in digital signal processing and communication systems. I also knew that I wanted to go to graduate school. I interviewed with Bell Laboratories in New Jersey and worked for them for three years, including a year in graduate school, which they paid for."

Jose E. Hernandez
Electrical Engineer
Lawrence Livermore
National Laboratory

When employers examine the resumes of engineering graduates, they look at more than their grade point averages (GPAs). They want to see evidence of leadership ability and a willingness to get involved and be a team player. Membership in a student chapter of a professional engineering society gives you a chance to demonstrate these characteristics. Three or four years of societal membership and one or two years service as an officer will help convince employers you have what they want. A combination of leadership activities and a strong GPA will really get an employer's attention.

Chapter 5 describes those professional engineering societies that maintain student chapters at most engineering schools.

4-2 LANDING A JOB, CO-OP, OR INTERNSHIP

Once you have earned your engineering degree, how do you get that first job? How about getting a co-op job or an internship? This section provides details on how you start by deciding what kind of organization you want to work for, the type to which you will be sending resumes. Your resume must be preceded by a cover letter that introduces yourself. The cover letter is what a potential employer reads first, so it better be flawless. The resume records your goals, lists significant work experience, and documents leadership qualities. If the employer likes both the cover letter and the resume, you may be asked to come in for an interview. If that goes well, you may have won your first job.

The process may sound simple, but it is not. Very few job applicants get it right. Most employers report amazement at the low quality of the cover letters and resumes they read, and the poor showing of many applicants in their interviews.

FOCUS ON

HOW I GOT MY FIRST JOB

"I had worked as a summer employee for four years—two years with Brown and Root and two with Schlumberger. My first full-time job was with Brown and Root as a construction engineer in the fabrication yard just outside of Houston."

Paul Olstad
Project Manager
Fluor Daniel

"I went to work for Boeing because my primary goals were staying in the Seattle area and working on large projects. (I'd spent three years working in the nuclear industry during the summer and found I enjoyed working on large projects.) I've been working at Boeing ten years now."

Clay Hess
Lead Engineer, Mechanical Hydraulic Systems Organization
Boeing Corporation

"After receiving my undergraduate degree from Boston University, I went to work for Digital where I designed the hardware for microprocessor-based optical disk drive controllers. After a few years, I took a sabbatical and received my master's degree from M.I.T., with my thesis work focused on VLSI circuit design. From there, I joined Digital's Semiconductor Group, where I've been ever since."

Sharon Britton
Principal Hardware Engineer
Digital Equipment Corporation

Choosing a Potential Employer

Before you write your resume, you should consider the type of company, agency, or institution you want to work for. Big? Small? Local? Government? International? Consulting? Research and development? Manufacturing? To help you make an informed decision, throughout your time in engineering school, you should read about business in general and organizations that employ engineers in particular. Read the business section of your local or regional newspaper. Read popular business magazines and the *Wall Street Journal*. Scan the want ads in the newspapers and the *Wall Street Journal* to see what opportunities exist.

As you approach graduation, you will have to choose the specific organizations from which you will seek employment. Of course, it helps if those organizations are seeking engineers. The primary resources for engineering employment ads are engineering journals. Chapter 5 includes a table of engineering societies, all of which publish journals in which organizations advertise when seeking engineers. Your engineering school and university library will have recent copies of journals that are relevant to the degrees they offer. You might also contact engineering alumni from your school and set up informational interviews to learn more about particular organizations or industries. Your deans should have a list of alumni who are willing to be contacted.

You should learn as much as possible about an organization before you send in a resume. This knowledge will help you write a quality cover letter and to perform well in an interview (both topics are covered later in this section). Most companies that hire engineers are public companies, with stock held and traded by the public. If a company you are interested in is a public company, there will be a wealth of information on it in the *Moody's* and *Standard & Poor's* directories. The *Thomas Register of American Manufacturers*, which is revised annually, is the Yellow Pages of American industry and will give you important information about companies involved in all fields of engineering. All three publications should be available at your university and public libraries.

Other sources of information on companies include the *Million Dollar Directory* and the *Business Directory*. When in doubt, ask the reference librarian, who is experienced at such searches and will quickly direct you to just the right place. If you have access to a business library, use it. Business libraries focus on information on business and companies.

FOCUS ON

HOW I GOT MY FIRST JOB

"I decided after receiving my bachelor's degree that I didn't really know as much as I thought I should. So I thought a master's would be a good idea.

I went to the University of Toronto, where I met a professor who was heavily involved in marine oil spills. He got me involved in the natural dispersion of oil slicks. Through his contacts, I ended up getting a job in Calgary, Alberta, which is the center of the Canadian oil industry. The job was with a company that was drilling in the Canadian arctic.

My job was as a research engineer, dealing with prevention and clean-up of oil spills. I spent four and a half years doing lots of experiments, simulating oil spills, testing various techniques, and handling oil in open water in the Beaufort Sea, north of Alaska."

Ian Buist
Vice-President
S. L. Ross Environmental
Research Ltd.

Resumes

The importance of having a quality resume cannot be overemphasized. Your resume is a sales brochure; its purpose is to land an interview. In the resume, you want to maximize your strong points and minimize your weaknesses. Have your completed resume reviewed by someone who has experience in writing resumes, perhaps a professor, engineer, or someone who regularly hires engineers. Most employers complain about the number of resumes they receive that have misspelled words and typographical errors. Such errors give a company an excuse for narrowing down the resumes they should seriously consider. If possible, attend a resume-writing workshop; most colleges offer one. Many word processing programs include one or more resume templates to help you get started.

The following is a recommended general structure for the resume of a new engineering graduate:

- **Name:** Center your name in bold, large type at the top of the page. Give your address, phone number, and e-mail address.
- **Objective:** State the type of position you seek, for example, "To work as a design engineer in microprocessor applications." Be reasonably specific.
- **Education:** List your degree. If you received it with honors, say so. State your overall GPA if it is 3.0 (out of 4.0) or higher. State your engineering GPA, especially if it is higher than your overall GPA. List the college courses you took that support your stated objective. Also list other courses that are relevant, for example, foreign language courses if you are applying for an overseas job. List your computer hardware and software expertise.
- **Awards:** List any awards and scholarships you received in college. Include scholastic, athletic, club, and service awards.
- **Experience:** List your work experience in reverse chronological order. Include jobs you held on campus, after school, and even while in high school. Employers like people who have worked, even when the work was unpaid or part time. It is particularly important to emphasize engineering-related jobs. Elaborate on particular achievements, such as your senior project, bringing together and leading a student committee, or helping a professor or employer install a computer network. Emphasize any computer skills you have, even if you are self-taught. Show evidence of your self-motivation, leadership abilities, and willingness to work hard.

FOCUS ON

INTERVIEWING

"I think, when I was being interviewed myself, the biggest mistake I made was being willing to be led too much by the different interviewers. I think I was trying to make a good impression in agreeing with people when they said, 'Doesn't it sound like fun?' or, 'Wouldn't you like to do that?'

Looking back, I think those interviewers probably ended up with a very erroneous impression of what I wanted to do.

Students should be frank and honest in an interview about what they want to do. I want people who have given some thought to what they want to do and who really want to come to work for me, doing the kind of stuff I do. They have to convince me that they want to work for me. That's what will make them stand out. There's a sea of people out there who want a job, so you have to stand out."

Tom Henderson

- **Activities**: Describe your activities in your school, community, and church. Stress membership and leadership roles in engineering societies.

Figure 4-1 gives an example of a good resume from a new engineering graduate.

Do not exaggerate on your resume. For example, you should not mention a club membership if you never attended any of its meetings, since the interviewer is likely to ask for details about your involvement. On the other hand, do not be modest. Your resume should be accurate and truthful, but it should also present you in your best light.

New graduates often feel that they have little to put in their resumes. But if you begin to focus now on the categories listed in this section, by the time you graduate you will have plenty to offer a prospective employer.

Cover Letters

Every resume you mail should be accompanied by a unique cover letter that will call attention to your resume. Before you write the cover letter, research the organization. Determine who should receive your resume, and address your cover letter to that person. The library resources previously described (in the section entitled "Choosing a Potential Employer") list company phone numbers you can call and ask to whom to send your resume. Avoid simply addressing your resume to human resources or the personnel department.

Let the reader know that you know something about the organization. Point out strengths in your resume that are applicable to the job you seek. Make it sound like this is the one and only resume you mailed out; never write an impersonal form letter. Keep it short, less than one page.

The following is a checklist for your cover letter:

- Is the letter unique, written expressly for the organization to which you will mail it?
- Is the letter addressed to a specific person?
- Do your opening sentences capture the reader's attention?
- Did you describe your major accomplishments, those that are relevant to the job you are seeking?
- Is the letter less than one page long and easy to read?
- Did you write in a clear, professional, business-like manner?
- Are you positive there are no grammatical or spelling errors?
- Is the letter clean, with dark, easy-to-read print?

Figure 4-2 gives an example of a good cover letter. Note that the letter is addressed to a specific person, starts by getting the reader's attention, indicates knowledge of the company, and emphasizes the writer's qualities that make him a good candidate for the job he seeks.

Figure 4-1
An Example Resume

John J. Smith

College Address: Permanent Address:
206 Anderson Hall   1234 Elm Street
Taft University   Sacramento, CA 95606
Bakersfield, CA 93309   (916) 555-1234
(805) 555-0805

OBJECTIVE
Obtain an entry-level civil engineering position with a large construction company
that offers opportunities for international employment.

EDUCATION
Bachelor of Science, Civil Engineering, Taft University, Bakersfield, CA
Overall GPA: 3.24
Engineering GPA: 3.40
Associate of Arts, Engineering, Bakersfield College, Bakersfield, CA
Overall GPA: 3.75
Minor in Spanish: 4 semesters of conversational Spanish.
Experienced in using PCs and UNIX workstations and in C programming.

AWARDS
1995   Bank of America Scholarship Award, $500.
1992 - 1996   Regent's Scholarship, covering half of tuition.
1992 - 1996   Dean's list every semester but one.

EXPERIENCE
Spring 1996   Senior Project, Taft University.
                  Designed and built wooden horse barn with storage for tack.

Summer, Fall
1995   Co-op, Jones Construction Co., Fresno, CA.
                  Assisted civil engineer in construction of large farm storage barn.
1993 - 1996   Lab Assistant, Taft University.
                  Managed the engineering school computer lab 3 to 9 hours/week.
1992 - 1996   Grader, Taft University.
                  Graded calculus, physics, and engineering homework.
1991 - 1994   Shift Manager, Pizza Hut, Bakersfield, CA.
                  Handled sales and managed 5 workers, part time during school
                  years, full time during summers.

ACTIVITIES
1994 - 1995   Chairperson, student chapter of the ASCE.
1994 - 1995   Member, Tau Beta Pi.
1993 - 1996   Member, ASCE.
1993 - 1995   Volunteer worker at county children's home.

REFERENCES
Available on request.

Figure 4-2
An Example Cover Letter

John J. Smith
1234 Elm Street
Sacramento, CA 95606
(916) 555-1234

April 13, 1996

Ms. Jane J. Doe
Human Resources Director
Magnum Construction, Inc.
Houston, TX 77707

Dear Ms. Doe:

Dr. Richard Turpin, professor of civil engineering at Taft University, suggested that I write you concerning a position with Magnum Construction. He said that at a recent ASCE conference in New Orleans you asked him to recommend graduating civil engineers who can help your company launch a project in Mexico. I am one of the students he recommended.

As you can see from the enclosed transcript, I have done well in my studies at Taft, and I expect to graduate with honors. I believe my strong GPA and my elective courses in construction and international studies make me a good candidate to meet your needs.

In addition to my academic strengths, I have a strong interest in working internationally. As a teenager I lived for a year in Zimbabwe, where my father worked as an electrical engineer. I've also traveled extensively throughout Africa and Europe. While I have not yet been to Mexico, I've taken three semesters of Spanish and am able to read and speak the language fairly well.

Ms. Doe, I look forward to speaking with you personally about my qualifications and how I can serve Magnum Construction. I'll phone your office on Tuesday, May 1, to inquire further about the position in Mexico. Perhaps we can arrange an interview at that time. Thank you in advance for your consideration.

Sincerely,

John J. Smith

Interviews

For many people the interview is the toughest part of the job-hunting process. Yet it is the most important part, so you must be well prepared. If you follow the guidelines outlined in this section, you can easily stand out as a top candidate in the eyes of a recruiter.

Recruiters will not expect you to have extensive experience or advanced engineering skills; they will be looking for potential. They will try to assess your grasp of basic engineering concepts. And they will look for positive personal traits such as self-confidence, flexibility, adaptability, and eagerness to learn.

The two most important points in an interview are the beginning and the end. The recruiter's initial impression of you is the most critical aspect of the interview. The momentum of a good start will help you relax and make the rest of the interview go more smoothly. To make a good first impression, dress well and be well groomed, be on time, and exhibit confidence. Introduce yourself, and shake the recruiter's hand with an eagerness that shows that there is no other place you would rather be.

Most recruiters mark the end of an interview with the question, "Do you have any questions about our organization?" Never say, "No." Come prepared with questions to ask. Ask questions that show an interest in the organization, not yourself. Ask questions that show that you have done some research on the organization. For example, "I know that XYZ Industries has offices in Atlanta. Are there opportunities for new engineers to work there?" or "I've read that XYZ Industries has grown 50 percent in the last two years. Do you expect such growth to continue?" During the initial interview it is generally not good form to ask, "How much will I get paid?" or "How many weeks of vacation will I get each year?"

You should walk into the interview knowing the following information, which usually can be found in any college library:

- The size of the organization
- What the organization does or manufactures
- In what cities, states, and countries the organization operates
- The direction the organization is headed and its potential for future growth
- How well the organization is doing in the marketplace and who is its competition

Do not ask the recruiter for this information. Having it before the interview will give you confidence. The recruiter will likely ask you what you know about the organization, but you can always volunteer the information in a natural way.

Questions you might ask during the interview include the following:

- What is the average time an engineer spends in project work before moving into project management?
- What is a typical career path in the area in which I would be working?
- What are my opportunities for travel?
- Does the organization have a formal continuing education program?

Summarizing, the following are a few good tips for doing well in an interview:

- Do some research on the company, agency, or organization. Know what it does and where it does it. Know where and how you might fit in.
- Dress in appropriate attire.
- Look the recruiter in the eye and shake hands with confidence when you enter the room.
- Do not sit down until invited to do so.
- Remember the recruiter's name when given it, and then use it when thanking the recruiter for the interview as you leave.
- Be enthusiastic, demonstrating a positive outlook on life and your career.
- Have prepared answers for common interview questions, such as, "Tell me about yourself," "What do you know about our organization?," "Why should I hire you?," and "Where do you plan to be five years from now?"
- During the interview, occasionally look the recruiter in the eye, but not so often that it makes the recruiter uncomfortable.
- When asked if you have any questions, ask relevant questions about the organization. Use your questions to demonstrate your knowledge of the organization.

You must enter the process of finding a job with reasonable expectations and established priorities. Be flexible, willing to work in another city, state, or even country. Be willing to work in a field other than your first choice, but do not appear desperate by being "willing to do anything." Do not make money your first priority. If possible, talk to some engineers who already work for the organization and get their advice. If you follow these simple rules, your name will be much more likely to rise to the top of any recruiter's list of potential employees.

4-3 SALARY AND EMPLOYMENT TRENDS

Trends in engineering employment have been fairly steady during the 1990s. Growth in employment has been about 4 percent per year, and salaries have been keeping pace with inflation. Figure 4-3 shows the trend in

FOCUS ON

RESEARCH

"I was part of a joint research project with Sandia National Labs in their combustion research facility (CRF). I took a sabbatical there, and we did a joint study working with some people who specialized in combustion engines. They were curious to see if they could get augmented heat transfer by impinging the exhaust, which is pulsating exhaust onto a flat plate.

The team members came in with very different strengths. One person's strength was theoretical, management, project planning; he was the leader of the project. Another expertise was in combustion and fluid mechanics. I came in with expertise in conductive heat transfer and external flows. We had a blast working together, setting up the experiment, finding what went wrong, and then rerunning it until all hours of the night. It was wonderful."

Pamela Eibeck
Mechanical Engineering Department Chair
Northern Arizona University

the granting of engineering bachelor's degrees over a ten-year period. As the figure indicates, the general trend is downward, indicating the possibility of a future shortage of engineers. A future shortage is also indicated by the College Placement Council (Bureau of Labor Statistics, 1995), which projected a 34 percent increase in engineering positions by the year 2005. Engineering is predicted to be one of America's top ten careers in the 21st century. Figure 4-4 shows average salary trends for all engineers. Figure 4-5 shows average 1993 salary offers according to engineering field.

Figure 4-3

Trends in Earned Engineering Bachelor's Degrees from 1985 to 1994 (*Source:* **Engineering & Technology Degrees 1994,** *published by the ASEE*)

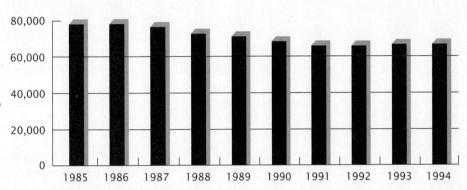

Figure 4-4

Average Salary Offers for Engineers Receiving Bachelor's Degrees from 1985 to 1993 (*Source:* **Statistical Abstract of the United States 1994, published by the U.S. Department of Commerce**)

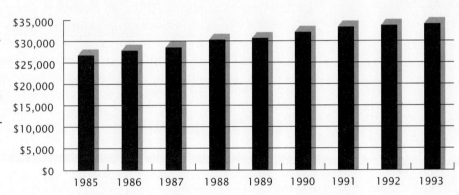

Figure 4-5

Average 1993 Salary Offers by Degree for Engineers Receiving Bachelor's Degrees (*Source:* **Statistical Abstract of the United States 1994, published by the U.S. Department of Commerce**)

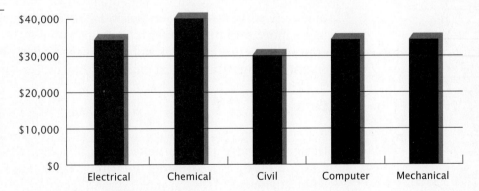

No one can confidently predict where the most jobs will be in three or four years, when you graduate. However, studies do indicate that there will be a shortage of engineers well into the 21st century. In the mid-1990s manufacturing, computer hardware and software, construction, environmental, and safety engineering are showing the greatest growth.[1] As automobile, aircraft, machine tools, food processing, chemicals, paper, and textile companies expand, the manufacturing sector shows the greatest potential for hiring engineers. Other strong areas include the petroleum, tire and rubber, and electronics industries. At the other end of the spectrum is the defense industry, which is scaling down while moving to adapt itself to the private industry market.

Prospective employers today look for engineering-related work experience, whether cooperative or summer internship. They also seek engineers with good interpersonal skills. They want engineers who can function productively as part of a team, who have strong communication skills, and who interface well with customers. Software skills are extremely important to most employers. Business knowledge also can be beneficial. Employers complain, however, that new engineers often want too much money, are unwilling to start at the bottom of a corporate structure, and are inflexible about relocating to a new part of the country or world.

Be aware that engineering job security may become an oxymoron as we enter the 21st century. Many companies are moving toward hiring temporary engineers from consulting firms, allowing them to hire engineers only as needed for specific projects. Although some people predict that the full-time engineering job may be disappearing, large companies that continuously develop engineering projects will still require the services of full-time engineers. Nonetheless, as a new engineer entering the job market, you should plan for an uncertain future and be prepared to be mobile and to stay current in your field.

4-4 CAREER OPPORTUNITIES

The typical engineering degree program will train you as a designer and enable you to function in a broad range of career fields. However, only a fraction of engineering graduates actually take jobs as designers straight out of school. Those with bachelors degrees often go into production, operations, or sales. Those with masters or doctorate degrees may take positions as researchers or teachers. Later in their careers, many engineers move into management. Some go into consulting. Your engineering education, with its mathematics, science, and humanities and social sciences requirements, should enable you to succeed in virtually any field.

Design

As previously stated, most engineering curricula emphasize design. Designers bridge the gap between an idea and the production line. First and foremost, as a designer you must be creative. You must also have solid analytical, decision-making, and problem-solving skills. You must have strong mathematical and scientific knowledge and have an understanding of economics, being aware of the costs involved in both the design and production of a product. You must be able to work under pressure, adhering to schedules while functioning effectively as a team member and demonstrating the communication skills that are necessary for the success of the team.

Most design positions demand that the engineer take his or her design through a process that results in a working product. Therefore, as a designer you can expect to be involved in development and testing. Development takes an idea from a set of design drawings to a working prototype or final product. It includes obtaining parts and materials and interacting with technical, construction, or other specially trained personnel. Once you have developed a prototype or final product, you must test it to ensure that it meets the original specifications.

Given the broad range of skills and talents required of a design engineer, the easiest path into this career is to get a masters degree or to work your way into it through time and experience in other engineering roles. Design work is exciting and rewarding as you guide ideas into reality.

Production

While a designer typically cooperates with other engineers who design the various parts of a complex system, the production engineer coordinates the activities of the people who build the systems designers create.

Experienced production engineers are often responsible for estimating the costs of production, determining, for example, the cost of building a bridge or manufacturing 10,000 cameras. To do cost estimation, they must be able to estimate accurately the cost of fabrication materials, using various manufacturing processes, and labor, as well as be able to estimate overhead and profit. As a production engineer you might be responsible for bidding on projects and then establishing the schedules and budgets for projects.

Production engineering is a career of great responsibility, bringing together and coordinating the human, material, and monetary resources that are necessary for the creation of a product.

Operations

As an operations engineer you would be responsible for the operation of a complex system, such as a manufacturing or processing plant. Another title for an operations engineer is plant engineer. The operations engineer ensures that large systems operate efficiently, safely, and legally. For example, the piping, boilers, turbines, generators, and monitoring equipment of any power plant, whether nuclear or fossil-fuel-fired, form a complex system that is typically managed by a team of operations engineers. Constant monitoring of the system is necessary. Maintenance schedules must be established and maintained. All state and federal regulations must be observed. The safety of plant personnel must be ensured.

An operations engineer must have a broad range of knowledge. A background in industrial engineering is most helpful, but knowledge of civil, electrical, mechanical, and chemical engineering, as well as economics and business law, can also be useful.

Operations engineering is a challenging career requiring broad knowledge, extensive communication skills, and the ability to work closely with people from many disciplines. The satisfaction of having the big picture and control over the way things are done can make this a rewarding career.

Sales

Many engineering graduates find interesting and lucrative careers in sales. Sales engineers are not salespeople. They do not knock on doors or seek sales orders. The sales engineer is a technical person who works with the sales staff, answering technical questions of potential customers and educating existing customers.

Many products created by engineers are extremely complex. A well-trained sales engineer can unravel that complexity for potential customers and convince them that a product will be an effective solution to their problem. As a sales engineer you would train the customers how to use the product and offer technical service. You would also provide product support, seeking ways to improve the products you sell.

You must be comfortable with people and have a pleasing personality to succeed as a sales engineer. A background in psychology, sociology, and human relations will also be useful. Sales engineering offers an interesting career that always involves new places and new faces and often involves significant travel.

Research

As a researcher you would apply basic scientific principles to the discovery of new knowledge that could help solve society's engineering-related problems. You must be particularly creative. You also must be patient since a researcher can work many months or even years on one problem.

FOCUS ON

CONSULTING

Joe Engel started his engineering education at Bakersfield Junior College in Bakersfield, California. He then double-majored at the University of California at Davis, receiving bachelor's degrees in mechanical and civil engineering. Today he is part owner of Engel & Company Engineers, an engineering consulting firm in Bakersfield, California. Joe Engel reflects on his 18-year consulting career:

"When I graduated from college, my dad, who owned his own consulting firm, talked me into joining the firm. Once I got out into the real world of jobs and people, I discovered I really didn't know a lot about the practical side of engineering. So I started out as basically a draftsman. But as time went on, I was taking jobs from start to finish. And that's what I've been doing ever since.

One of my biggest challenges as a consulting engineer is not only to figure out the engineering solution but also to deal with the personalities involved. I try to encourage a certain level of agreement among all members of a team so that the project will move ahead. I learned a lot from watching my father in meetings, seeing how he dealt with others. These were things I could never have learned in the classroom.

I cannot imagine a more fulfilling occupa-

Students often assume they have to be professors to be researchers, but many government agencies and most large companies also employ researchers. Some federal agencies, such as the National Science Foundation (NSF), pay researchers to seek advancements in virtually all fields of engineering and science. Others, such as NASA, seek knowledge in more narrow areas. Although many researchers who receive support from the NSF, NASA, and other government agencies are professors, a large number work directly for federal, state, or local governments.

Both large and small companies perform research to improve their products and gain a larger segment of the market in which they compete. Most companies—for example, those in the transportation field—also perform research to find ways to increase the safety of their products. A good research and development (R&D) department can help a company remain competitive in our increasingly technological society.

Whether you seek employment as a researcher for a university, the government, or industry, the education requirements are basically the same. An advanced degree, usually a Ph.D., is a must, with a strong background in mathematics and science. You must also have a strong desire to seek and discover new knowledge. The element of discovery, coupled with solving real-world problems, can make research a particularly satisfying career.

Management

Managers manage people, materials, money, and time. They oversee the activities of the employees they supervise and help guarantee the success of projects by establishing budget and time constraints.

Managers spend much of their time reading, writing, and meeting with people. They review proposals, contracts, and specifications and keep up with new laws. They read and write letters, memos, progress reports, and personnel evaluations. Often managers have to do the bulk of their read-

tion. When you own your own company, your position is defined in the way that you want to define it. You can work 80 hours per week, or you can work 10. In addition to the freedom, you get the satisfaction of putting together projects any way you want. And you don't have to deal with the in-fighting and jockeying for position that you see in large organizations.

Unfortunately, I think a lot of young engineers try to go into consulting too soon, without enough experience. They just get their licenses and think they can hang out a shingle and practice engineering. They make a mistake, I think.

My advice to the student who wants to become a consultant is to work with an older engineer for about ten years, doing the work he or she doesn't want to do. You will end up solving so many engineering problems that when you do go out on your own, you will be confident and able to provide good service.

In any event, I think newcomers should try to start their careers in an established consulting firm. Then they can move into a partnership, becoming a consultant in a more natural way, with less risk."

Joe Engel

ing and writing at home because so much of their time at the office is taken up with meetings—with assistants, engineers, their own managers, clients, and government officials, just to name a few.

Engineers tend to be promoted from designer to project engineer. As a new project engineer, you will likely be put in charge of a project or part of a project, with several engineers under your supervision. In this position you will oversee the design work of others on the project and will likely still do some design work yourself. The goals of your projects may or may not be initially clear; how to achieve the goals will be up to you. It will be your responsibility to bring the project in on schedule and within budget.

Over time, a competent project engineer will be given responsibility for managing increasingly large projects, or multiple projects involving the supervision of other project engineers. Depending on the size of the company, a project engineer may eventually be promoted to an executive management level, such as vice-president. Vice-presidents typically oversee entire sectors of their companies, including design, production, or operations.

Engineers are often viewed as having the problem-solving skills required to manage large companies. As a result, many companies that are not involved in engineering nevertheless hire engineering graduates and place them directly into management positions, usually after some training. Companies that deal with technology are even more likely to prefer to have engineers in their management positions. An engineer with an MBA is an even more desirable candidate.

FOCUS ON

TEACHING

Pamela Eibeck received her B.S., M.S., and Ph.D. degrees in mechanical engineering from Stanford University, then began her career as a university professor. Eibeck describes why she finds teaching so rewarding:

"I remember teaching a lab one time. We were measuring drag on different-shaped vehicles. I had the students read a research paper that was written by a GM engineer. The engineer happened to be at Stanford at the time for a conference. As I heard him explain his research, I remember being so excited seeing the meaningfulness of what the students were learning in school and its relationship to what is actually happening in industry.

When I was teaching at Berkeley, I was the campus Principle Investigator for a coalition of teachers from several universities whose task was to investigate innovative teaching. I discovered how much I liked having the big picture, having a sense that I could influence a strategic plan of a program.

As chair of the Mechanical Engineering Department at NAU I'll be teaching two courses each semester. But I'll also be working on some curriculum reform, adding some new emphases to mechanical engineering, and hiring some new faculty. So I'm very much involved in teaching, in a larger, envisionary way. Yet I am still involved with the students. I would not enjoy my job if I were not involved with the students."

Pamela Eibeck
Mechanical Engineering
Department Chair
Northern Arizona
University

Consulting

Engineers who move into consulting typically do so after they have worked several years in industry and have developed the experience and contacts they need to strike out on their own. Frequently two or more engineers leave a company together to form their own consulting firm. Some engineers join a consulting firm straight out of school. Larger consulting firms may employ more than 100 engineers.

In addition to engineering experience, a consultant needs business acumen to succeed. In fact, most consulting practices that fail do so because of a lack of knowledge of, or preparation for, the business end of the practice. Newcomers often face problems managing employees, dealing with administrative details, marketing their services, and handling cash flow, as clients are sometimes slow in paying for consulting services. Therefore, it is important to have ample cash reserves when starting out. Some say those reserves should be enough to last 6 to 12 months without a paying client.

Consulting is an inviting option to those who prefer to work for themselves and are willing to accept business risks that could result in great financial reward. Today the consulting field is doing well and appears to be poised for growth. Many companies are moving toward hiring fewer full-time engineers and toward hiring part-time consultants. This trend is likely to continue as companies strive to be as flexible as possible in order to compete.

International Opportunities

International jobs have much to offer an engineer. Learning a new culture can be interesting and exciting. Although learning a new language may not

FOCUS ON

INTERNATIONAL ENGINEERING

Paul Olstad received his B.A. from Harvard University and M.S. from Rice University, in civil engineering. After graduation, he worked for Brown and Root and the Burroughs Corporation for several years before joining Fluor Daniel, where he began an international career that has lasted 14 years. Olstad describes how he came to work overseas for Fluor Daniel, and the rewards of international assignments:

"I was sent to Malaysia to assist the construction of an on-shore oil terminal. They wanted somebody who knew process, control systems, electrical, civil structural, and mechanical. Of course, I didn't know all those things, but I learned pretty fast and did what needed to be done.

I'd been interested in an overseas assignment for a long time. Prior to that I had been on a few business trips and had a two-month assignment in Singapore and a two-month assignment in Denmark, but Malaysia was my first permanent assignment. It was supposed to last for five months, but it stretched into 51 weeks.

After Malaysia, I got an assignment to Saudi Arabia. I went there in January and brought my children over in June, after they were out of school. We spent just over two years in Saudi Arabia the first time. While there, I was project manager for a number of off-shore platform projects.

After two years, a decrease in construction activity in Saudi Arabia had virtually shut down our office there. So Fluor Daniel brought me back to Houston. Two years later, things be necessary, often it is an option that the engineer should pursue. In addition to higher pay, an international career offers interesting travel and cultural experiences in a variety of locations.

Although jobs in the less popular overseas locations can pay as much as double what you could get in the United States, you should think carefully before accepting an international position. The weather may not be to your liking. Living conditions may be less than ideal, and your travel within the host nation may be restricted. Some countries are inhospitable, or even off limits, to minorities and women. Be sure you are ready to stay for awhile when you go, since a commitment to stay overseas for at least two years is a common requirement.

Teaching

Most engineering disciplines offer numerous teaching opportunities. Although you may be able to find a teaching position at a technical school that requires that you have only an undergraduate degree, most institutions require an advanced degree. Teaching at a community college typically requires a masters degree, and teaching at a college or university requires that you earn a Ph.D. Most schools prefer that instructors have some experience as practicing engineers; however, this is not the case with most universities because they want professors to concentrate on research.

Most teachers enjoy the challenge of increasing others' knowledge and understanding, the stimulating interaction with students and colleagues, and the excitement of research. Add to these benefits a relatively flexible schedule and the variety of two new semesters every year, and it is clear that teaching can be a rewarding career.

had improved in Saudi Arabia, and I was asked if I wanted to go back and start up our office there again. Of course, I went.

I spent about two years helping build the Saudi Arabia office from nothing to about 100 people. Then we were awarded a major program.

A week later, Saddam Hussein invaded Kuwait. About a week after the invasion, my company evacuated everyone to Jeddah or out of Saudi Arabia, and I returned to Houston for two weeks. It was an interesting experience.

"The pay you can expect from an overseas assignment depends on the country. Almost any overseas assignment includes some increase in salary or benefits and also some assistance in accommodations. Just as a reference, for Saudi Arabia, our standard work week is 48 hours instead of 40 hours, so the engineers get a 20 percent pay increase for that. And we get another 15 percent increase as a foreign incentive. And we are provided housing, and an area increase for the cost-of-living differential.

Another nice thing about working overseas is the opportunity to travel, which we did whenever we got our home leaves. Our children were in school in Europe, so we went to Europe often. We also went to the Far East, spent Christmas in Australia, and did a lot of things that we could never have done otherwise."

Paul Olstad
Project Manager
Fluor Daniel

4-5   MAPPING A CAREER

Being in control of your career takes planning that should start now. Develop both short- and long-term goals. Generally your long-term goals determine your short-term goals. For example, if your long-term goal is to have your own civil engineering consulting firm, your short-terms goals should include taking courses in civil engineering, economics, and business management and getting a civil engineering degree. If you plan to move into management eventually, you should consider taking some business and accounting classes; you may even want to earn an M.B.A. If you prefer to stay involved in technical work your entire career, you may want to pursue a masters degree or a Ph.D. in engineering.

Once you have joined a company, continue to develop short- and long-term goals. Ask yourself what you want to be doing a year from now as well as ten years from now. Plan how best to achieve your goals.

Chances are, as you get started as a novice engineer, you will be given relatively simple assignments. You will be expected to be a team player and to learn from the more experienced engineers. In a year or two you will likely be assigned larger projects, lasting several months or even a year or two. Within several years you may be promoted to project engineer. If you stay on a typical career track, you will eventually be put in charge of increasingly larger projects with more engineers for you to supervise.

Most companies allow you to pursue a career in a purely technical role if you decide that a management role is not for you. The income and promotion opportunities may be greater for managers in some companies; however, many companies now offer parallel career paths for those engineers who prefer a technical career to a management career. As a technical engineer you could move into positions in which you do design work as part of a team that you would also manage. Or, if you have an advanced degree, you may move into research and other more scientifically oriented work.

Whatever your career goals, you should keep them continually in mind and consciously guide your career. Grasp opportunities as they present themselves. Always be a team player. Be flexible, adaptable, and a willing learner. Change is about the only constant in an engineering career, and education is your best ticket to ensuring you keep up.

SUMMARY

It is important for you to start planning your engineering career now. The goal of this chapter was to show you how. In addition to pursuing your engineering degree, you should join one or more engineering societies and get involved in their activities. If your school has a cooperative education program, consider taking time off from your studies for a co-op job, or find work as a summer intern at an engineering company. Remember the advice offered here on choosing a potential employer and on writing resumes and cover letters. The library reference desk is a resource that you will find useful throughout your entire career.

The salary and employment trends in engineering indicate you have wisely chosen to be an engineer. The prospects are bright as we enter the

21st century. Although most engineers are trained to be designers, engineering offers many careers other than design work, including production, operations, sales, research, management, consulting, and teaching. You can be in control of your career, but it takes planning and effort.

Problems

1. Interview a practicing engineer to learn what role his or her engineering studies have played in job and career development. Based on your interview, what advice would you offer to your fellow students about their college education?

2. Interview a professional who is not an engineer but has an engineering background (such as a doctor, patent lawyer, small business owner) to learn what role his or her engineering studies have played in job and career development.

3. List the engineering-related work experience you presently possess, if any, and the experience you plan to have upon graduation.

4. Attend a student professional society meeting on your campus in the discipline that most interests you. Write a report about the meeting, indicating the number of attendees, the person or persons who ran the meeting, the issues discussed, the plans made, and how you would contribute to the society if you were a member.

5. For what specific organizations would you like to work during your college years to gain engineering experience? Investigate what types of opportunities are available at those organizations.

6. Using Figure 4-1 as a model, write your resume as you would like it to appear ten years from now. What actions will you have to take to ensure that this resume becomes a reality?

7. Write a cover letter that would accompany your resume. Address it to the organization for which you would most like to work. Use Figure 4-2 as a model.

8. How would you answer the following common interview questions: What are your short-term and long-term goals? What contribution would you make to this organization? Where do you plan to be five years from now?

9. Of the engineering careers discussed in this chapter, which do you think you would enjoy most? Least? State the reasons for your choices.

10. Find out which industries hire the type of engineer you feel you are best suited to be, using information from journals, newspapers, and your campus career and placement centers.

5 The Successful Engineering Career

"**M**y primary responsibility is with hazardous waste. I'm always trying to find ways to save money, to reduce the amount of waste we generate," says Rebecca O'Dell, an environmental engineer for the U.S. General Services Administration (GSA). "The GSA has all sorts of paints and sealants and chemical products that they ship to other federal agencies. One problem is that they always have too much in stock, to make sure they never have zero stock. And so some of the material goes out of shelf life.

"Before I came to the GSA," O'Dell continues, "everything that was out of shelf life or no longer met specs was going out as hazardous waste. It was

costing about $800,000 a year for hazardous waste disposal when I came on board. I said, 'This needs to be changed. Not all of this is hazardous waste. There are other things we can do with this material. We can give it away, we can reuse it, we can recycle it.' My primary goal was to see most of it recycled.

"So I put a hazardous waste contract in place. (They didn't have one in place when I arrived.) And I put a recycling contract in place. Then we opened what we call a SAVE store. It's a value store where other federal facilities or qualifying agencies like the Boy Scouts or Girl Scouts or homeless shelters can come and get things for free or for a dollar an item. Basically, it's a re-use center."

Rebecca O'Dell's recycling efforts have resulted in substantial savings for the GSA. "Recycling hazardous materials always costs money," she says. "But it costs about 20 percent of the cost of hazardous waste disposal. We pay probably $30,000 per year now for hazardous waste disposal. That's a savings of $770,000 a year."

Continuing education is a fact of life for environmental engineers, O'Dell notes: "There is absolutely no way you can know all of the regulations. But what you can know is where to look, and you have to know how things relate." To keep up with changing regulations, she takes "a lot of training classes, all the time—basically about new laws, how we are going to respond, what kind of reports we have to do, and how we are going to comply with the laws. That's one thing students should know: When you get out of school, you aren't finished."

I asked O'Dell if she's had to deal with any ethical issues as an environmental engineer. She replied, "When I did my first audit, my boss was really nervous about going up the chain. He said, 'Are you sure you want to be as harsh as you've been?' I told him, 'I don't know any other way to be. If they are unhappy with the audit, I've given them options and recommendations on how to correct their problems and I'll be more than happy to assist with that.' But I said, 'There is no way I can change any recommendations I've made.' Once people realize that you are going to stand by what you believe in, you aren't going to be involved in any ethical problems."

INTRODUCTION

Chapter 3 outlined the basic skills and education required to become a practicing engineer, and Chapter 4 described a number of career opportunities open to engineers. This chapter describes the activities that help ensure a successful career for a practicing engineer. In the 1990s the successful engineer is one who plans for the future, makes informed career decisions, stays technologically and professionally current, stays involved in engineering societies, is ethical, environmentally conscious, and keeps a global view of the engineering profession.

5-1   CAREER DECISIONS

As stated in Chapter 4, a major part of the process of making informed career decisions involves keeping your eye on your future and maintaining short- and long-term goals. This process should begin early in your engineering education and continue throughout the course of your career. Because of the many career paths open to engineers, you will likely work in a number of different positions and work environments during your career. This characteristic of diversity is one reason why engineering is such an interesting and rewarding career.

Career Paths

Your first career decision upon graduation will be choosing the type of organization with which you will seek employment. Organizations that hire engineers include corporations; local, state, and federal government agencies; research organizations; and educational institutions. Corporations tend to offer higher wages than most government agencies, but they may also offer less job security. Research organizations generally require graduate degrees but offer state-of-the-art work. Teaching has its own special rewards, such as student interaction and flexible hours, but usually typically offers less pay than a corporate position.

You must also decide whether to work for a small or large organization. Small organizations can give you a chance to get a broader picture of engineering and to become a generalist. They often promote engineers more quickly into positions of responsibility. A disadvantage of working for small organizations, however, is that they often have fewer resources and less advanced facilities. They may not have the mentors you will need early in your career to help you develop as an engineer and make wise career decisions.

Large organizations may have extensive resources and advanced facilities, but they may also have rigid rules about the promotion, design work, and even dress of engineers. They may be more likely to entice you into being a specialist in a field that may be short-lived. However, large organizations offer larger projects and more varied opportunities.

As you begin your engineering practice with a specific organization, one of your first career decisions will be whether to specialize or become a generalist. Of course, most engineers fall between the two extremes of knowing much about a very narrow subject and knowing little about a very broad range of subjects.

Specialists try to be on the cutting edge of their fields and must work particularly hard to stay current. Organizations seek them out for their expertise and are willing to pay high consultanting fees. Specialists must be aware, however, that their areas of specialization may become obsolete as technology evolves.

Generalists have the comfort of a broad range of knowledge, allowing them to move easily into many different areas of engineering. However, most employers seek engineers with at least some specialized knowledge.

Once you have completed one or more projects at your first organization, you may want to try something completely new. For example, you may want to move to an organization that is more or less technologically advanced than your present employer. Many high-tech organizations offer exciting opportunities. They can also be more volatile, hiring engineers when they get a big government contract, only to lay them off a year or two later. Low-tech organizations can be less volatile, offering more stable but less state-of-the-art employment.

Early in your career you should observe managers in your organization and begin to think about whether or not you would want to move into management at some point. This is an issue that you may consider and prepare for in college, but you cannot really make the decision until you have actually worked on engineering projects and seen management in action. If you do decide on a management career path, you may want to take some courses in business and management if you have not already done so (see Section 5-2). You will need to prepare for a career of administration, dealing with people, and making sound technical decisions. Let it be known in your organization that you want to be a manager and accept positions of increasing responsibility when offered.

FOCUS ON

LIFELONG LEARNING

"I had no appreciation of how critical reading is to one's success in a profession. But as I look back, I recall that my Dad always read after dinner. After watching the news, he'd pull out his journals and sit and read about the latest in chemistry."

Pamela Eibeck
Mechanical Engineering
Department Chair
Northern Arizona
University

"I visit the local bookstores on a regular basis, always looking for the most recent books in my field. I also read technical journals at our LLNL library. And I go to trade shows to see the latest technology commercially available."

Jose E. Hernandez
Electrical Engineer
Lawrence Livermore
National Laboratory

"I think that staying current in your field is very important. For me, as an industrial engineer, it's not so much an issue of keeping up with engineering technology as it is staying current as a professional person. I've stayed very much involved in different professional societies such as ASME (American Society of Mechanical Engineers) and ASQC (American Society for Quality Control). They are the best source for information on seminars and special development programs, which are critical to the success of any professional career."

Greg Raco
Project Manager
Ocean Spray, Inc.

Getting Licensed

The purpose of registration laws for engineers, like those for doctors and lawyers, is to protect the public. All 50 states and the District of Columbia have engineering registration laws. Although the laws vary from state to state, all states require licenses for those engineers who are responsible for public projects, such as highways, bridges, and power distribution systems. Some states also have laws requiring licenses for consultants, expert witnesses in court cases, and engineers in certain high-level corporate positions.

In general, engineers who are employed by private companies need not be licensed as long as they are not responsible for public projects. However, it is usually to your advantage to get licensed, even if the law or your organization does not require you to do so. Having a license is viewed by most engineers as proof of competence and as an extension of a sound education. It is a positive addition to any resume.

Interestingly, licenses are usually required for engineers who design static systems, but often those responsible for dynamic systems need not be licensed. Therefore, nationwide, far more civil engineers are licensed than automotive or aeronautical engineers. Yet dynamic systems, such as automobiles and aircraft, have an equal or greater potential for doing harm than static systems, such as highways and bridges.

The legal process for becoming a licensed engineer varies among the states. In general, however, it involves four steps:

1. Earn an engineering degree from an ABET-accredited or approved institution.
2. Pass the Fundamentals of Engineering Examination. This exam is often called the EIT exam because passing it gives you the title of engineer in training. An eight-hour exam, it tests your knowledge of general engineering subjects, such as mathematics, physics, chemistry, mechanics, electrical science, thermal science, economics, and ethics. Whatever your plans may be, you can and should take the EIT exam during your last year of engineering school. You will probably never be more prepared to pass it, and doing so will enhance your resume and reassure you that you have built a solid technical foundation on which to build your career.
3. Complete four years of engineering practice as an EIT and obtain documentation of this experience from licensed engineers.
4. Pass the Principles and Practice Examination. This is an eight-hour exam that tests your knowledge of your engineering specialty.

In most states the license you are granted is valid for life unless revoked for legal cause. States generally require license renewal every one or more years, a process for which a fee is charged. Given the rapidly changing nature of engineering, some states now require licensed engineers to show evidence of professional activity or continuing education to remain licensed.

5-2 STAYING CURRENT

Ask experienced engineers how many of the tools they now use are tools that they learned to use in school, and they will say, "Not many." For engineers now in their 50s, the most important tool they learned to use in school was the slide rule. In the 1970s they moved to hand-held calculators. For more complex problems, they used computers and programming languages such as Fortran. More recently engineers have moved to commercial computer-based tools, such as spreadsheets, MATLAB, MathCAD, and Mathematica.

Today all engineers are relying increasingly on personal computers and workstations and specialized engineering application software, such as computer-aided design, modeling, simulation, and manufacturing tools. Due to the pervasive use of computers, most engineering design processes differ radically today from what they were a mere decade ago. Although the fundamental mathematical and scientific tools you are now learning in engineering school will remain useful throughout your career, you too will have to adapt to new tools as they are developed in the future.

Lifelong Learning

Lifelong learning is a fact of life for engineers. If you plan to remain active in engineering over any extended period of time, you must commit yourself now to a life of continuing education. The education you receive during your 4 or 5 years in college cannot adequately meet your needs during the following 40 or 50 years.

FOCUS ON

ETHICAL ISSUES

"Achieving a balance between safety and cost is one of the most difficult and interesting ethical issues a civil engineer must deal with day in and day out. It is hard to explain to the public that there is no such thing as a perfect engineering design. Even the best engineering work involves the possibility of failure. I think the biggest obligation of civil engineers is to provide engineering designs of the greatest safety at a reasonable cost."

Guoming Lin
Geotechnical Engineer
S&ME, Inc.

"In general, for a small consulting firm, the biggest ethical dilemma is cost versus performance. We make money by making proposals and getting those bids accepted. More and more today, it's price that counts more than experience or professionalism. You're constantly put in situations where clients expect good, professional work, but want it for virtually no cost. And that, to me, is the biggest ethical issue.

"You just have to be very honest with your clients. In my work it is particularly difficult. In

R&D you don't know how long it is going to take you to find an answer, if there is one. There's an awful lot of guesswork, but experience goes into trying to find an appropriate price at the beginning of the job that will get you to the end successfully."

Ian Buist
Vice-President
S. L. Ross Environmental
Research Ltd.

Early in your career you will likely wish you had taken both more practical and more fundamental courses. Later, as you move into positions of greater responsibility, you may wish you had taken more business and management courses. Late in life you might want to know more about the humanities. Throughout your career you will want to know more about your own field of engineering and about the fields of those with whom you work.

Fortunately a practicing engineer has many opportunities for continuing education. Most engineers continue their education primarily by reading books and journals. Useful journals include scholarly publications of engineering and other technical societies, as well as the more easily understood trade journals. Trade journals are independent magazines that are often free to engineers because of the large number of advertisements they contain. These advertisements can be great sources of information on the products and activities of the companies in your field. There are also the popular scientific journals, such as *Science* and *Scientific American,* and the practical magazines, such as *Popular Mechanics* and *Popular Electronics.* Most of these journals and magazines review and advertise recent books in the fields they cover. You can order these books directly from their publishers. Of course, your public library or university library will likely have a selection of books on engineering and other technical subjects in your field. Internet discussion groups also are a great source of the latest developments in all areas of technology.

Many organizations encourage and subsidize continuing education. They offer short on-site courses, tuition reimbursement for courses taken at universities or from private organizations or engineering societies, and funds for book purchases. Some organizations grant leaves of absence and financial aid for the pursuit of graduate degrees.

FOCUS ON

ETHICAL ISSUES

"We are working on a project with a company called Medtronics that builds pacemakers. These are software intensive systems. So as a software engineer on these projects, not only do you want to build good software, but you also have to realize that your failure might cause someone's death. Because the world is becoming so software intensive, ethics is more in the forefront."

Grady Brooch
Chief Scientist
Rational Software
Corporation

"I think that society has a view of technology that is erroneous. I think engineers owe it to society to tell it over and over again that technology will not solve all of its problems. Technology is a tool. But societies are built of something more than technology.

"Technology is a two-edged sword. For example, it is technology that has made us efficient at waging war. Even the technology that allows us to construct a building is a two-edged sword. That technology can be used to store munitions as easily as food.

"The point is that we must be careful how we use our technology. When you give someone a power saw, you hand it to the person and you say, 'You're going to build your house with this power saw. But if you're not careful, you're going to cut off your fingers.'"

Joe Engel
Structural Engineer
Engel & Company
Engineers

Most engineering societies are dedicated to the continuing education of their members. Among other things, all societies offer journals and other publications that contain information on recent research, descriptions of professional developments, and tutorial papers. For example, the Institute of Electrical and Electronics Engineers (IEEE) publishes *Spectrum*, the American Society of Mechanical Engineers (ASME) publishes *Mechanical Engineering Magazine,* and the American Society of Civil Engineers (ASCE) publishes *Civil Engineering.* These and other societies publish hundreds of other periodicals as well. Engineering societies also sponsor conferences that offer half- or full-day tutorials covering current topics. For those who prefer not to travel, many societies offer satellite seminars, videotapes, and home study courses. Some employers subsidize or pay for society memberships and publications.

As you advance in your career, you may find your public education system a better educational source than your employer or engineering society. All community colleges offer low-cost courses, such as business, personnel management, and economics, that can help you move into project and personnel management positions. Many colleges and universities offer graduate engineering and MBA degrees. Universities that are near centers of technology, such as Silicon Valley in California, offer interactive video courses that let you communicate with an instructor without leaving your company or organization site.

Engineering Societies

Engineering societies are a very important asset in your effort to stay current in your field. As previously noted, one of their major concerns is education. There are more than 400 engineering societies and related engineering groups. The oldest of these organizations is the ASCE, founded in 1852. The largest engineering society, IEEE, was founded in 1884 and currently has more than 300,000 members worldwide. Nearly all engineering societies concern themselves with one, often small, segment of the engineering profession. Exceptions include the American Associa-

FOCUS ON

THE GREEN ENGINEER

"All engineers have a tremendous obligation to protect the public. As professionals, the public depends on us to make the right decisions and also to protect the environment. If we can do things in an environmentally sensible way, I think that's important. And we should not waste resources."

Ian Buist
Vice-President
S. L. Ross Environmental
Research Ltd.

"As engineers involved in designing, developing, and manufacturing products, we should be very concerned with the consequences and impacts such products will have on our environment. There is only one earth with a finite amount of resources, many of them critical to human existence. We all need to take responsibility to ensure that in the long run we are

making this world a better place in which to live."

Jose E. Hernandez
Electrical Engineer
Lawrence Livermore
National Laboratory

tion of Engineering Societies (AAES), the National Academy of Engineering (NAE), the Accreditation Board for Engineering and Technology (ABET), and the National Society of Professional Engineers (NSPE), which were created for the benefit of all types of engineers.

In addition to sponsoring conferences, most engineering societies involve themselves in the same basic activities, including the following:

- Enhancing the standing of engineers within society
- Maintaining and improving standards of engineering education
- Engaging in government lobbying activities that benefit their members
- Encouraging engineering research, design, development, and production activities
- Providing information on engineering research and related topics in journals and other publications

The two most visible activities of engineering societies are publishing technical journals and organizing technical conferences. For example, IEEE organizes approximately 300 conferences each year and publishes nearly 100 monthly journals.

Society journals are sources of not only the latest information in specific engineering fields but also job advertisements. The conferences offer opportunities for networking with other engineers from all over the country and even the world. Many engineers find new jobs through the informal meetings they have with other engineers at conferences. Table 5-1 gives the names and phone numbers of societies you can contact to obtain information on conferences and other educational and networking opportunities.

Engineering societies establish industry standards in areas such as construction, electronics, and ethics. They also recognize exceptional accomplishment by granting awards, often including monetary grants, to both

FOCUS ON

THE 21ST CENTURY ENGINEER

"I believe that in today's society it is important to have a specialty, that is, to be very good at something. Also, due to job insecurity and more world-wide competition, you need to be versatile enough to adjust to the demands of the marketplace. In other words, you have to be a good and fast learner and be willing to become an expert in more than one thing."

Jose E. Hernandez
Electrical Engineer
Lawrence Livermore
National Laboratory

"You've got to pick your specialty, carve out your niche, and make sure you're the best engineer you can be in your specialty. And I can't think of an engineering profession in which you don't need to be mobile. That's a part of the profession. You go where the jobs are."

Ian Buist
Vice President
S. L. Ross Environmental
Research Ltd.

"The global economy is challenging practicing engineers to work more efficiently. Competing on a worldwide scale will demand a strong academic background, knowledge of different disciplines, and an understanding of nonnative cultures."

Guoming Lin
Geotechnical Engineer
S&ME, Inc.

students and professional engineers. Most societies offer scholarships to their more promising student members.

As a student member of a professional engineering society, you will likely be subsidized by practicing members, so your dues will remain low while you are in school. Among other things, dues are used to lobby Congress on your behalf and pay for the publications and newsletters most society members receive every month.

There are too many professional engineering societies to list them all here. Table 5-1 lists the more popular national societies, those that often support student chapters at engineering schools. (Check our Addison-Wesley Toolkit web site at http://www.aw.com/cseng/toolkit/ for the e-mail addresses of these societies.) A complete list of engineering societies can be found in the *Encyclopedia of Associations*, which includes addresses and phone numbers. Ask for it at any library's reference desk.

Table 5-1 Professional Engineering Societies

Field	Society	Phone
Aerospace	American Society of Aeronautical Engineers	(412) 776-4841
Agricultural	American Society of Agricultural Engineers	(616) 429-0300
Biomedical	Biomedical Engineering Society	(310) 618-9322
Ceramics	National Institute of Ceramics Engineers	(614) 890-4700
Chemical	American Institute of Chemical Engineers	(212) 705-7338
Civil	American Society of Civil Engineers	(202) 789-2200
Computer	Institute of Electrical and Electronics Engineers Association for Computing Machinery	(212) 705-7900 (212) 869-7440
Electrical	Institute of Electrical and Electronics Engineers	(212) 705-7900
Environmental	Institute of Environmental Sciences American Academy of Environmental Engineers	(708) 255-1561 (410) 266-3311
Industrial	Institute of Industrial Engineers	(404) 449-0460
Mechanical	American Society of Mechanical Engineers	(212) 705-7722
Power	Institute of Electrical and Electronics Engineers	(212) 705-7900
Manufacturing	Society of Manufacturing Engineers	(313) 271-1500
Materials	Materials Engineering Institute	(216) 338-5151
Mining	Society of Mining Engineers	(303) 973-9550
Nuclear	American Nuclear Society	(708) 352-6611
Ocean	The Society of Naval Architects & Marine Engineers	(201) 798-4800
Petroleum	Society of Petroleum Engineers	(214) 952-9393
Systems	National Council on Systems Engineering	(408) 773-8701

Several engineering societies are geared toward minority groups, including the American Society of Black Engineers [(703) 549-2207], the Society of Hispanic Professional Engineers [(213) 725-3970], and the Society of Women Engineers [(212) 509-9577]. Minority and female engineers may want to join those societies that best serve their needs. (It is a good idea to join the mainstream societies as well.) If you qualify for membership in an engineering honor society, you should also maintain that membership. If the society of your choice does not have a student chapter at your school, you can contact it about helping establish one.

The importance of staying current in the knowledge, technology, and practice of your profession cannot be overemphasized. In this rapidly changing environment, those who fail to keep up will not succeed. As long as you are an engineer, you must also be a student and remain involved in the evolution of your profession.

5-3 ETHICS

Engineers frequently have to make tough ethical decisions that involve a wide range of issues, from balancing cost and safety to addressing the environmental impact of their designs. In certain cases, poor engineering decisions can lead to loss of life. Examples include the Challenger explosion, the Chernobyl nuclear accident, the Union Carbide plant gas leak in Bhopal, India, and the Ford Pintos that exploded upon rear impact. Clearly, ethics are central to the engineering design process. For this reason most engineering groups have developed a code of ethics to guide their members.

Ethical Issues

One of the most common ethical issues an engineer faces is that of providing a quality product at reasonable cost. Sometimes clients want more than they are willing to pay for. The engineer must decide whether to cut corners in a design to save money or time, or simply to refuse to do the job. On the flip side of the coin is the issue of wasting the client's money by overspecifying a design to ensure it meets certain specifications. Spend too little to satisfy the customer, and strength or safety may be compromised. Spend too much, wasting the customer's money, and you may not get any more jobs as other firms out-bid you on future work.

The ethical engineer designs not only for safety and low production cost, but also for lower future costs. For example, some of the costs of a machine design not seen up front include the machine's life expectancy, speed of operation, power efficiency, and required number of operators. These are costs that a client may not completely consider. The ethical engineer is qualified to find the correct balance of present and future costs and is self-confident enough to know that the resulting product will be safe. Good communication skills come into play as you convince clients or management that they can trust your expertise. However, at some time in your career, you may experience a situation in which a client or manager cannot be convinced. To maintain an ethical posture in such a case, you may have to refuse to bid on a project or refuse to approve a design.

Other ethical issues are those of confidentiality and conflict of interest. As an engineer you may be privy to your employer's trade secrets, patent applications, and copyrights. This information may be on a computer system that allows you to make copies easily. Such information should obviously stay within company walls. If you leave a company, you should return or erase all data, whether hard copy or on computer disks. Engineers who change employers must also ensure that their knowledge creates no conflicts of interest. For example, they should not work on a product that will compete with a former employer's product if they have signed an agreement that specifies that they not do so for a certain period of time.

Whistle-blowing is a controversial issue that is sometimes faced by engineers, especially when contrasted with their obligation to maintain the confidentiality of colleagues and employers. If you become aware of criminal activity, such as industrial espionage or bribery, you have an obligation to report it. If faced with a situation in which you believe your organization or someone within it is creating a design that might jeopardize public safety, you will have to decide whether or not to act. Of course, you should attempt to resolve any problem within the organization first. If that attempt is unsuccessful, whistle-blowing may be your only alternative. However whistle-blowing does have consequences, including the loss of your colleagues' respect or friendship to the loss of your job.

A basic issue in ethics is simply telling the truth. When you give an oral presentation or write a report, be sure that what you say is the complete truth. Omit no relevant information. Be aware that your managers will always want to hear only good news but that you cannot always give it to them. A good case study of political pressure affecting engineer performance is the Challenger disaster. The rocket engine manufacturer, Morton Thiokol International, was in the process of contract negotiations with NASA. The Shuttle project was behind schedule, and Congressional support was waning. The result was a situation in which the engineers were feeling peer and management pressure to "get the product to market" while overlooking obvious safety issues.

If you become a manager, you will be responsible for not only your own performance but also the performance of others. Encourage those working for you to be honest in their reporting. When things go wrong, it is not acceptable to lay the blame on a junior engineer. It will be your job to see that things are done correctly. Read your engineers' reports critically, and be ready to ask questions.

Ultimately the ethical decisions you will make as an engineer will depend on the situations in which you find yourself. Many engineering schools are now including discussions of ethical responsibility in their curricula. These discussions are typically included both in freshman introduction to engineering courses and in senior project courses. If they are not, you should do your own research on the subject. A good approach is to read case studies and decide what you would do if you were involved in the cases studied. See our web site address, http://www.aw.com/cseng/toolkit/, for exercises and case studies on the subject of ethics.

The "Green" Engineer

The industrial revolution of the 19th century marked the beginning of today's environmental pollution problems. Over the past 100 years, industry and the internal combustion engine have generated uncountable millions of tons of pollutants. Until recently the philosophy in dealing with the problem could be stated as, "The solution to pollution is dilution." The ill effects of contaminating products were minimized by merely mixing the products with air, water, or land. Smoke stacks were made taller to emit exhausts into faster moving air, chemicals were dumped into the oceans because it was thought that their capacity to absorb waste was limitless, and landfills were made increasingly large as users ignored the fact that much of what was placed in the landfills was not biodegradable.

It has now become clear that the earth's resources are not limitless and that the environment cannot forever sustain high levels of pollution and other forms of degradation. Today engineers must factor environmental costs into the equation when designing new products and technologies.

In the past, pollution control was considered after the fact, at the end of a process. Companies added scrubbers to smoke stacks or catalytic converters to cars. Today, however, environmental considerations must be factored into the entire design process. Those concerned realize that pollution control must begin at the beginning. Design engineers must design products that can be cleanly manufactured, maintained, and ultimately recycled or salvaged. Materials engineers must design materials that do not include, or do not require in their production, chemicals that harm the environment. Production engineers must design processes to use more environmentally friendly chemicals. Sales engineers must convince clients that they will benefit from purchasing environmentally sound products.

Progress is being made. Many organizations are rethinking their engineering and production processes. For example, many newspapers have substituted soy-based ink for oil-based ink in color printing. Some industries have substituted water-based cleaning agents for chlorinated solvents. The Air Force now removes old paint from airplane fuselages using a spray of plastic pellets rather than chemical solvents. Many organizations have adopted improved chemical storage, handling, and recycling practices.

While there are many government regulations and organizational guidelines that force engineers to deal with ecological issues, engineers should be leaders, not followers, in this crucial area. The engineering profession still has a good public image. To maintain that image and remain self-regulating, engineering professionals must uphold high ethical standards, hold public safety as the highest priority, and have respect for the environment and the conservation of its resources.

Code of Ethics of Engineers

The Accreditation Board for Engineering and Technology (ABET), the accrediting body for engineering schools in the United States, developed the code of ethics of engineers as seen in Figure 5-1.

The ABET code of ethics is just one example; every major engineering discipline has developed its own unique code. You should obtain a copy of your discipline's code of ethics from the student chapter or national office of your engineering society. You might find it interesting to compare this code of ethics with ABET's or those of other disciplines.

Figure 5-1
**Code of Ethics
of Engineers**

The Fundamental Principles

Engineers uphold and advance the integrity, honor, and dignity of the engineering profession by:

1. using their knowledge and skill for the enhancement of human welfare;
2. being honest and impartial, and serving with fidelity the public, their employers and clients;
3. striving to increase the competence and prestige of the engineering profession; and
4. supporting the professional and technical societies of their disciplines.

The Fundamental Canons

1. Engineers shall hold paramount the safety, health, and welfare of the public in the performance of their professional duties.
2. Engineers shall perform services only in areas of their competence.
3. Engineers shall issue public statements only in an objective and truthful manner.
4. Engineers shall act in professional matters for each employer or client as faithful agents or trustees, and shall avoid conflicts of interest.
5. Engineers shall build their professional reputation on the merit of their services and shall not compete unfairly with others.
6. Engineers shall act in such a manner as to uphold and enhance the honor, integrity, and dignity of the profession.
7. Engineers shall continue their professional development throughout their careers and shall provide opportunities for the professional development of those engineers under their supervision.

5-4   THE 21ST CENTURY ENGINEER

To survive in the 21st century, you will have to be adaptable. The world is changing at an increasingly rapid rate, and changing with it will take some effort. You will have to be flexible and willing to stretch yourself and accept challenges. You will have to develop and maintain a global perspective throughout your career. That perspective will include being knowledgeable about the world around you and willing to work overseas and in teams with engineers from other countries. It will also include having a sensitivity toward other cultures and recognition of the diverse requirements of global customers. Throughout your career you will have to remain vigilant, always looking to and planning for the future.

As international trade barriers drop and international transportation and communications systems expand, the world is becoming more interdependent. This factor and the increasing industrialization of less developed countries are encouraging American corporations to expand internationally. American companies are not only marketing their products in other countries but also performing research, establishing manufacturing sites, and providing engineering services in other countries.

Nearly all large organizations that hire engineers now have international operations. To succeed in these organizations, you will need a strong education in the histories, cultures, and customs of other countries. Speaking a foreign language and traveling broadly will also help you succeed in an international organization.

SUMMARY

To be a successful engineer over the long term, you will have to maintain a high level of energy and involvement in your profession throughout your career. As this chapter indicates, the successful engineer

- Makes informed career decisions
- Stays current in his or her field
- Is ethical
- Is sensitive to environmental issues
- Is adaptable and flexible
- Is globally minded

The 21st century will offer new and difficult challenges to engineers. The problems will be many, including dealing with a shifting global economy, increasing competition from other countries, a diverse multicultural work force, shrinking product life cycles, and constraints established by legal, regulatory, and environmental standards. But the opportunities are many as our increasingly technological world demands more skilled and talented engineers.

Problems

1. Develop a set of specifications for the ideal organization you would like to work for after you graduate. Research companies to find which ones best meet your specifications. Resources you can refer to include *100 Best Companies to Work For, Moody's,* Standard & Poor's, the *Thomas Register of American Manufacturers, Million Dollar Directory,* and the *Business Directory.*

2. Would you prefer a long-term technical or management career? State the reasons for your choice.

3. Interview a successful practicing engineer to find out what he or she did to become successful. Write a short report on your findings.

4. Attend a society meeting of professional engineers. Before and after the meeting, discuss with attendees their discipline and the qualifications required for success (specific college subjects, additional degrees, licensing, and so on). Based on this information, write a prescription for success as an engineer in that discipline.

5. Research how Henry Ford financed his first automobile manufacturing plant and how Steve Jobs and Steve Wozniak financed their first Apple computer assembly plant. Report your findings. If you had a great idea for a new product, how would you finance its production?

6. Were you to practice engineering in the city in which you now live, what resources would be available to you for staying current in your field? Include lifelong learning opportunities in your list. For ideas, check the library, local college catalogs, and the publications of professional engineering societies.

7. Attend a society meeting of professional engineers. Ask two attend-ees what contributions they make to the society and what benefits they receive from membership. Write a short report on your find-ings.

8. What do you think is the most serious ethical issue facing your engi-neering school? What are your reasons for choosing this issue?

9. If you knew a group of your fellow students had devised and were going to implement a way to cheat on an upcoming exam, would you report the students to the professor? Why or why not? Would it make a difference if the professor graded on a curve, so that the cheating would lower your grade?

10. Consider the ramifications of a federal law that would ban all inter-nal combustion engines. Make two lists, one stating the positive effects and the other stating the negative effects.

11. Investigate and report on how a specific object or product, such as aluminum, glass, newspaper, or automobile tires, is recycled.

12. Read a book or article that makes predictions about events or issues that will affect the practice of engineering in the 21st century. Sum-marize its predictions. In what way might these events or issues affect the engineering profession?

Index

GLOUCESTER MASSACHUSETTS

123 embroidery

EASY PROJECTS FOR ELEGANT LIVING

ROCKPORT PUBLISHERS

ellen moore johnson

First published in the United States of America under the title *The Embroidered Home* by Rockport Publishers, Inc.
33 Commercial Street
Gloucester, Massachusetts 01930-5089
Telephone: (978) 282-9590
Facsimile: (978) 283-2742
www.rockpub.com

We have made every effort to ensure that the instructions, illustrations, and diagrams are accurate and complete. We cannot, however, be responsible for human error, typographical mistakes, or variations in individual work.

ISBN 1-56496-475-2

10 9 8 7 6 5 4 3 2 1

Design and Layout: Leslie Haimes
Cover Design: Laura McFadden
Cover Images: Bobbie Bush Photography, www.bobbiebush.com
Photography: Bobbie Bush Photography, www.bobbiebush.com
Illustrations: Pages 15-21, Judy Love. All other illustrations and diagrams by Lorraine Dey Studio.

Printed in China.

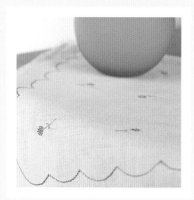

Introduction

It seems we've always been searching for new ways to bring style to our living quarters. Our fascination with transforming an ordinary dwelling into a unique personal oasis can be traced all the way back to the ancient Egyptians, the first civilization known to have embellished handwoven linen fabrics with embroidery. The desire for beautiful surroundings remains deeply embedded in our consciousness today. It is our goal to help you in your quest to create a comfortable and cozy place to live, to help you turn your dwelling into a true home.

Have you ever thought about the factors that govern the manner in which you decorate your home? Your family, your age, your career, your friends, even your hobbies—all of these things, and more—influence the decisions you make about your home's furnishings and overall decor. Many different styles result from the merging of our diverse backgrounds, but one simple common thread continues to tie us all together: We all love to create, to make things with our hands.

Perhaps you're yearning to decorate a box with beautiful stitchery. Maybe memories of cooking and chatting with Mom in the kitchen have sparked a desire to embroider a tablecloth. Whether you embroider a gift of elegant linen bed sheets to help a young couple set up housekeeping or you whip up a whimsical shower curtain for your daughter's first apartment, you are adding beauty and inspiration to the lives of those you love. The needlework you create today will be passed down to future generations and become cherished possessions that tie together a family's history. Cherished treasures that bring happiness. Heirlooms.

We invite you to join us as we embark upon our tour of The Embroidered Home. It doesn't matter whether you're a novice needleworker, a veteran stitcher, or someone who tried her hand at embroidery once upon a time but who hasn't picked up a needle in many years. Let's take a peek ...

SECTION ONE: Embroidery Basics

Before you can begin embroidering, you must become acquainted with materials and equipment appropriate to your project. The array of fabrics, fibers, and tools available to today's needle artist is vast. Knowing which ones to choose will make your embroidery go smoother and ensure the longevity of your finished piece.

This section covers the obvious embroidery requirements—needle, stitching fiber, and background fabric—as well as auxiliary supplies, from scissors and stilettos to tracing vellum and marking chalk. You'll learn how to prepare a pattern for embroidery, transfer the design to the background fabric, and secure the fabric in a hoop or frame. Refer to this section as needed when working on an individual project. We've also diagrammed a few of the basic embroidery stitches featured in the projects to help you get started.

SELECTING NEEDLES

Many types of needles are used in embroidery, and one of the embroiderer's essential skills is the ability to select the appropriate needle(s) for a project. The needle should be slightly thicker than the thread, so that it will lay open a sufficiently large passage in the fabric for the thread to pass through. Needle size is indicated by number; the lower the number, the larger the needle.

Crewel Needles

Crewel needles, commonly referred to as embroidery needles, are sized #1 through #10. They are of medium length with a long, oval eye and a short point. Here are some recommended uses:

#7 or #8	beginning shadow work; beginning Point de Paris
#9	shadow work; Point de Paris (pinstitch); stem stitch; detached chain stitch; granitos; backstitch; cutwork
#10	shadow work (this small needle size is perfect for rendering the tiny, even stitches necessary for smooth curves); hemstitch; satin stitch; granitos; stem stitch; drawn thread work

Crewel needles #1 through #8 will accommodate three to six strands of stranded cotton, silk, rayon, coton à broder (twisted, nondivisible thread), broder medicis (fine wool thread), #8 perle cotton, #12 perle cotton, and fine metallics; crewel needles #9 and #10 will accommodate one or two strands of cotton, silk, or rayon.

Sharps

Sharps are general-purpose needles, sized #1 through #12. They are of medium length with a small, round eye and a sharp point. Here are some recommended uses:

#10	beginning bullions; hems; general sewing (e.g., fine hand sewing)
#11	small, tight bullions
#12	small, tight bullions; satin-stitched dots; tiny whip stitches for hand appliqué (superior for achieving fine detail)

Sharps #7 through #9 will accommodate two or three strands of stranded cotton, silk, or rayon; #10 through #12 will accommodate one or two strands of stranded cotton, silk, or rayon.

Milliner's Needles

Milliner's needles, also known as "straw" needles, are sized #1 through #15. They are extra long with a round eye and a sharp point. The eye is no wider than the shaft, allowing for easy manipulation of bullion wraps. Here are some recommended uses:

#8	beginning long bullions
#10	long bullions

Milliner's needles #1 through #4 will accommodate four to six strands of stranded cotton, silk, rayon, coton à broder (twisted, nondivisible thread), #8 perle cotton, #12 perle cotton, and metallics. Milliner's needles #9 through #11 will accommodate one or two strands of stranded cotton, silk, or rayon.

Tapestry Needles

The distinguishing feature of the tapestry needle is its blunt, rounded tip. This special feature enables the needle to slip easily between the warp and weft threads without piercing or snagging the fabric. Tapestry needles are sized #13 through #28. They are of medium length and have a long, oval eye. Here are some recommended uses:

#26—Fil Tiré (punch work); silk ribbon embroidery (on delicate fabrics such as handkerchief linen and swiss batiste); pulled and drawn thread embroidery; cross-stitch; petit point

#28—Fil Tiré (punch work); shadow work (the small needle size is perfect for rendering tiny, even stitches)

Tapestry needles #26 and #28 will accommodate multiple strands of stranded cotton, silk, rayon, fine metallics, and broder medicis (fine wool thread) or single strands of silk ribbon, coton à broder (twisted, nondivisible thread), #8 perle cotton, and #12 perle cotton.

Chenille Needles

Chenille needles are excellent for working with thicker threads and silk ribbon. They are sized #14 through #26. Chenille needles are of medium length with a long, wide oval eye and a sharp point. Here are some recommended uses:

#22	beginning silk ribbon embroidery
#24	silk ribbon embroidery (the sharp point proves quite useful when working on a densely woven fabric by diminishing the fraying of the ribbon)

Chenille needles #22 and #24 will accommodate tapestry wool, crewel wool, six strands of stranded cotton, #3 perle cotton, #5 perle cotton, thick silk, heavy metallics, and silk ribbon.

Betweens

Betweens, also referred to as quilting needles, are sized #1 through #12. Quite short in length, betweens have a small, round eye and a sharp point. Their small size is very good for rendering short stitches quickly and accurately. Here are some recommended uses:

#7	French knots; outline stitch
#10	running stitch; backstitch, beginning quilting; appliqué

Betweens #7 and #10 will accommodate floche à broder beautifully. Also recommended are one or two strands of cotton, silk, or rayon.

SELECTING FIBERS

The wide array of fibers available in today's marketplace makes thread selection more exciting than ever, but it can also create confusion for the embroiderer. The descriptions of fibers given here will help you appreciate their different attributes and uses.

Generally, you should match the fiber content to the background fabric. Use linen, cotton, or mercerized threads on linen or cotton fabrics, silk threads on silk fabrics, and wool threads on wool fabrics. The finished pieces will wash better because the fibers are spun to the same tension and have the same pull. When stitching items that won't be laundered frequently, you can relax the rule a bit.

Threads that fade or bleed can ruin a project, so it is very important to verify colorfastness. Even though many manufacturers label their products as colorfast, it is always wise to test any questionable threads before you begin stitching.

Some threads are particularly difficult to manipulate and tend to fray easily. Metallics, rayon, and silk fall into this category. If you choose these fibers, use short lengths—about 12" (30 cm)—to make them more manageable and to help them retain their luster.

Six-Strand Cotton Embroidery Floss

Cotton floss is the most commonly used embroidery thread. Its six mercerized two-ply strands can be separated to create different thicknesses of thread in the needle. Strands of different colors can be used together for added color depth and richness. Cotton floss is available in hundreds of colors and has a silky sheen. It is appropriate for all types of embroidery.

Overdyed Floss

Overdyed floss is usually a particular shade of six-strand cotton floss that has been dyed with additional colors to create a watercolor effect. The colors flow together in a pleasingly subtle manner, making this thread particularly useful in stitching flowers, leaves, and water.

Floche à Broder

Floche à broder is a mercerized five-ply, single-strand cotton thread that is available in over seventy colors. It has a silky sheen, and the individual strands are thicker than those in six-stranded cotton floss. Floche à broder lies very smoothly against the fabric, making it a wonderful choice for satin stitch and shadow work embroidery.

Coton à Broder

Coton à broder is a twisted, nondivisible mercerized cotton thread that is available in over one hundred colors. This versatile thread has a slight sheen and is especially good for cutwork, pulled thread, whitework, and monogramming.

Perle Cotton

Perle cotton is a mercerized, twisted, nondivisible thread. It is available in hundreds of colors and four different size gauges: #3 (the heaviest), #5, #8, and #12. Perle cotton's lustrous corded finish is good for creating textures. Recommended uses include canvaswork, needlepoint, and hardanger.

Broder Medicis

Broder medicis, also known as crewel wool, is a fine, two-ply, nondivisible yarn that can be used singly or in multiple strands. Broder medicis is the most commonly used wool thread for surface embroidery. It is available in more than one hundred colors and is quite suitable for needlepoint, crewel, and general embroidery.

Linen

Linen is a very strong, nondivisible thread. Its textured finish lends a natural effect. Primarily used to stitch on linen fabric, linen thread is available in both solid and overdyed colors.

Silk Ribbon

Silk ribbon has a lustrous finish and is favored for silk ribbon embroidery. It is available in over one hundred solid and overdyed colors and comes in four widths: 2 mm, 4 mm, 7 mm, and 13 mm. The 2 mm ribbon is excellent for small projects such as jewelry and for stitching small flowers and buds. The 4 mm ribbon is the easiest to manipulate, the most commonly used, and available in the widest array of colors. The 7 mm ribbon is useful for larger or more dramatic stitches. The 13 mm ribbon—the most expensive and the most difficult to find—works up into bold, luxuriant flowers and can be used to cover large areas quickly. Since silk ribbon frays easily, short 12" (30 cm) lengths should be used.

Silk Embroidery Floss

Silk embroidery floss has multiple strands that, like cotton floss, can be separated to create different thicknesses of thread. The individual silk strands are similar in thickness to those of cotton floss. Silk embroidery floss has a very high sheen and is available in hundreds of colors. The spectacular luster of silk makes it desirable for almost any type of embroidery. Caution is advised, however, because silk snags easily and frays quickly.

Rayon Embroidery Floss

Rayon embroidery floss is a six-strand, divisible mercerized synthetic fiber. It has a very high sheen, which makes it readily interchangeable with silk floss. Rayon floss is also usually less expensive—a bonus when cost is a factor. The disadvantages of rayon floss are that it snags easily and can be difficult to manage. The embroiderer who is willing to tackle these obstacles with patience and diligence should be well pleased with the end results, because rayon floss can provide an exceptionally striking accent to any piece of stitching.

Metallic Threads

Metallic threads are available in a multitude of colors and weights. Some metallics are actually wrapped sections of metallic plastic. These brilliant fibers add sparkle to your stitching, making them a wonderful choice for holiday projects. Due to the widely varying characteristics of metallic threads, each one should be evaluated individually.

SELECTING FABRICS

Fabrics for embroidery can be divided into two categories: evenweave and plain weave. Evenweave fabrics are those which have the same number of warp and weft threads per square inch. The exact number of threads per inch is referred to as the "count." Evenweave fabrics in durable fibers are particularly suitable for table linens. "Plain weave" refers to fabrics that are not evenly woven. Examples include dress and furnishing fabrics in both smooth and textured varieties, such as satins, slub silks, velvets, and heavy wools.

There are several factors to consider when determining the suitability of a fabric for a particular project. Will the article be used daily and laundered frequently? Do you require ease of cleaning and durability along with a dainty appearance? Or,

perhaps the embroidery will be purely decorative and framed under glass. When decorative pieces will not be laundered, the fabric selection can focus more upon aesthetics and less upon practicality. The needle artist can concentrate on choosing a fabric that works with the embroidery to convey a particular mood or feeling.

A final question is whether the piece is intended as a future heirloom and, as such, requires special consideration. An example of this type of article is a ringbearer's pillow—it must be fashioned from materials that are elegant yet able to withstand a child's inevitable touching and poking. Again, the question of washability arises. In all likelihood, such a piece will become soiled with handling and will require cleaning, particularly if the family plans to pass it on for use by future generations. A fabric that must fulfill so many requirements has to be selected carefully. While each situation is different, it is advisable to remember the advantages of using washable fabrics when soiling is a possibility.

Also be sure the fabric quality is worthy of the time you will spend on the embroidery. The importance of using good-quality fabrics cannot be overemphasized. Fabrics can be woven from man-made materials, natural fibers, or a combination of the two. For keepsake or heirloom pieces, natural fiber fabrics are preferable. Natural fiber textiles will endure for many generations when properly maintained. Here's a look at four fiber types you might use.

Cotton

Cotton is one of the most abundantly available fabrics today. Four kinds of cotton plants provide the raw materials for textile manufacturers. Sea Island cotton plants produce the finest-quality fibers, followed by Egyptian cotton plants, American Upland varieties and, finally, Asiatic plants. The majority of cotton fabric is manufactured from American Upland cotton varieties. Cotton yarn can be woven into many different kinds of fabrics. Batiste, broadcloth, cambric, crepe, drill, gingham, huck, lawn, nainsook, organdy, percale, piqué, poplin, sateen, swiss, and velveteen are but a few examples of cotton's versatility.

Linen

Linen yarn and cloth are produced from fibers obtained from the flax plant. Flax that has been harvested late in the summer produces the best-quality linen. The majority of the world's linen is produced in Europe. Belgium, France, and Ireland have become famous for their fine-quality linen fabrics. Cambric, damask, huck, and lawn are a few examples of linen fabrics.

Silk

Silk, one of the strongest natural fibers, is produced by silkworms. Each silkworm in its lifetime produces between 500 and 1,300 yards of fiber. The raw silk is reeled off the cocoons, twisted tightly into skeins, and shipped to manufacturers of silk fabric. Silk is very smooth and elastic. Dirt does not easily cling to it, making it one of the most useful materials of the textile world. The different ways of weaving and finishing silk provide a wide variety of finished materials: crepe, moiré, piqué, poplin, satin, shantung, and velvet, to name a few.

Wool

Wool yarn and cloth are produced from the sheared coats of sheep, goats, and members of the camel family. Most wool is obtained from Merino sheep flocks, but the African and Asian camel, alpaca, angora goat, cashmere goat, llama, and vicuna provide textile manufacturers with other special kinds of wool. Wool is primarily used for articles of clothing because it acts as a natural insulator. Shielding the body from varying outside temperatures, wool keeps a person cool in the summer as well as warm in the winter. Broadcloth, cashmere, crepe, and flannel are examples of wool cloth.

EMBROIDERY FRAMES

There are two advantages to using an embroidery frame. One advantage is that the work remains clean and unrumpled. Another is the uniformity of stitch tension. There is no better way to perfect your stitching technique than by working on a piece of fabric that is held taut and secure in a frame.

Embroidery frames are available in different shapes and sizes. The two most commonly used are ring frames, or hoops, and scroll frames. The mounting of embroidery fabric into a frame is called "dressing the frame." The most important point to remember when dressing any type of frame is that the fabric must be stretched with the warp and weft threads running at right angles to one another to prevent distortion.

Hoops can be made of wood, plastic, or metal. They are available in many sizes, the most common being 4", 5", 6", 8", and 10" (10 cm, 13 cm, 15 cm, 20 cm, and 25 cm). Round frames in larger sizes are often available with a table clamp attachment that includes screws for adjusting the angle and height of the frame. Round frames with table or floor stands are yet another option.

Small hoops are suitable for embroidery designs that can fit within the ring diameter. This way, the entire design can be stitched without having to reposition the fabric in the frame. The best type of round frame has an adjustable screw on the outer ring, allowing for even distribution of tension across the surface of the hooped fabric. Spring-type round frames do not produce this result and are not advised. There's a lot of tension on the fabric when it is in the hoop. To avoid damaging delicate fabrics, wrap the inner ring with cotton tape before dressing the frame.

Scroll frames, usually made of wood, are extremely versatile. They have two rollers, each with a cloth strip attached, to which the edge of the embroidery fabric is stitched. The rollers are held apart and parallel by side bars. Pegs in the side boards can be moved to adjust the frame to the appropriate size. The rollers may be turned to move the fabric or regulate the tension. Scroll frames can accommodate small or large pieces of work. They may be purchased with table or floor stands, or used alone as a lap frame.

THE EMBROIDERER'S WORKBASKET

While every needle artist's workbasket is unique, there are some items no embroiderer should be without.

- Embroidery scissors (for cutting threads)
- Dressmaker's shears (for cutting fabric)
- Paper scissors
- Tape measure
- Ruler
- Silk pins and pincushion
- Fine-line water-soluble pen (to mark embroidery design lines on fabric)
- Soft lead (#2) pencil (to mark embroidery design lines on fabric)

- Tailor's chalk (to mark sewing lines on fabric)
- Tracing paper (lightweight and heavy vellum)
- Graph paper
- Pounce powder and pad
- Tracing wheel
- Dressmaker's carbon
- An assortment of needles
- Thimble(s)
- Fingershield
- Stiletto
- Needlecase
- Thread organizer
- Assorted small hoops
- Cloth tape (to wrap the inside ring of round frames)

Scissors

The temptation to use the most conveniently located pair of scissors can be great, especially when only a few threads need snipping. Resist the urge to engage in scissor swapping. Scissors that are used to cut paper become dull very quickly, and dull scissors will damage intricate stitching. The blades will "chew" the fabric and threads instead of neatly cutting and trimming them.

Two pairs of sharp scissors are needed for embroidery—a pair of 6" (15 cm) or 8" (20 cm) dressmaker's shears for cutting fabric and a small pair of embroidery scissors, usually about 3" (8 cm) in length, for trimming threads and cutting intricate shapes or holes. Embroidery scissors should have very sharp, narrow, pointed blades with perfectly closing tips. All sewing and needlework scissors must be kept sharpened, and each pair should be stored in its own sheath when not in use. Scissors purchased without a protective sheath should be stored in a scissor case.

Other types of scissors that may prove helpful in various situations are lace scissors and blunt pocket scissors. Lace scissors are small scissors with a 1/4" (5 mm) duckbill at the ends. The special duckbill feature allows you to trim away excess fabric without accidentally snipping the lace. Blunt pocket scissors are used for general trimming. They are small with rounded tips, but the blades are very sharp.

Good scissors are costly, but the investment is worthwhile. Stitching is much more enjoyable when done with appropriate and properly maintained equipment.

Pins

Only premium-quality silk pins are acceptable when working with fine fabrics. Poor-quality pins have dull points and rough shafts. These defects can mar beautiful cloth and ruin hours of work. Many poor-quality pins have plastic heads that will melt onto the material when touched with a hot iron. Premium-quality pins have very sharp points, smooth shafts, and metal or glass ball heads. These pins glide through the fabric and do not melt if pressed over with a hot iron. Always avoid sewing over pins with a sewing machine to eliminate the possibility of breaking the needle or damaging the machine.

Thimbles and Fingershields

Thimbles and fingershields, used properly, can prevent the misery of sore fingers so often endured by avid stitchers. Thimbles are caps or guards used to protect the middle finger of the stitching hand when pushing a needle through fabric. In a properly fitted thimble, the pad of the finger should almost touch the bottom. On the outside, thimbles must be deeply pitted to catch the eye of the needle; a smooth finish prevents roughening of the thread. An embroiderer working on a fixed frame will require two thimbles—one for the hand beneath the frame and the other for the hand that works on top.

Fingershields are plastic or leather forms that fit over the index finger of the nonstitching hand. Fingershields prevent nicks and scratches caused by the sewing motion of a needle.

Stilettos and Awls

Stilettos and awls are pointed instruments that are used for forming holes in fabric without cutting or breaking the fibers. The point is used to spread the fibers apart. A stiletto is wide at the base, or handle end, and tapers gradually to a point. An awl is the same diameter from the base to the beginning of the pointed tip. While both tools are used in stitching eyelets, stilettos are more common.

PROPER LIGHTING

Eye fatigue is a common problem for needle artists. A very bright light and a magnifying glass will help alleviate the symptoms associated with overstrain. Fluorescent lamps are preferable because they emit "cool" light. Nonprescription reading glasses or a magnifying glass that can be hung around the neck are both viable magnification options. A wonderful invention and worthwhile investment is the magnifier lamp. This fully adjustable swing-arm lamp features a magnifying glass encircled by a fluorescent bulb.

Many optometrists recommend a stitching regimen that consists of thirty-minute work sessions interspersed with ten-minute breaks. This practice, coupled with the use of appropriate lighting and magnification devices, should eliminate virtually all eye fatigue.

TRANSFERRING THE DESIGN TO FABRIC

There are three popular methods for transferring embroidery designs to fabric: direct tracing, dressmaker's carbon, and pricking and pouncing. With each method, accuracy is extremely important. Sloppily transferred designs can result in imprecisely placed stitches that distort the original pattern. There is nothing more disheartening than spending hours of stitching time only to be disappointed with the final product because the design appears lopsided or does not align properly on the straight grain of the fabric. Readying your project for stitching can seem like a dull chore, particularly when you can barely wait to thread your needle, but doing it properly is essential.

Preparing the Fabric

First, preshrink the fabric by laundering it according to the manufacturer's guidelines. You can skip the preshrinking step only if the finished piece will not be washed. Apply spray starch over the entire surface of all fabrics except silk until slightly damp. After the cloth has absorbed the starch, iron it until dry. For silk fabrics, simply press with a warm iron. If you are embroidering on a cut piece of fabric, zigzag the outer edges on a sewing machine, or overcast them by hand, to prevent fraying. This step is obviously not necessary when embroidering a purchased article, such as a kitchen towel or pillowcase.

Before actually transferring an embroidery design to fabric, you must determine its placement. The most common practice is to center the design on the fabric. To find the center of the fabric quickly, fold it in quarters and mark the intersection of the two crease lines with a water-soluble pen or by inserting a straight pin. Use the same folding method to locate the center of your paper template, and mark it with a colored pen.

Direct Tracing Transfer

The direct tracing transfer method is easy to do. It is most successful with light-colored and delicate fabrics.

1. Trace or photocopy the embroidery design onto heavy vellum to create a template. If needed, outline the design with a black felt-tip pen so that it will be darker and easier to see when the fabric is placed over it.

2. Tape the design template to a smooth, flat, light-colored surface, such as a countertop, a solid-surface cutting board, or a piece of sturdy white corrugated cardboard. Place the fabric to be embroidered on top of the template (you should be able to see the design through the fabric). Make sure the design is properly centered and/or aligned, and then tape down the fabric.

3. Using a soft lead (#2) pencil or a fine-line water-soluble pen, trace the design onto the fabric. Work slowly and carefully; an accurately transferred design always results in a more beautifully stitched piece. If the design lines are difficult to read through the fabric, try taping the template and fabric to a light box or a large sunny window for tracing instead.

Dressmaker's Carbon Transfer

The dressmaker's carbon transfer method requires the use of a special fabric carbon and a stylus. Fabric carbons are available in several different colors, so select one that will stand out when it is applied to the material. This method is suitable for densely woven and dark-colored fabrics.

1. Trace or photocopy the embroidery design onto heavy vellum to create a template.

2. Tape the fabric to a clean, flat surface, such as a countertop. Place the design template on top of the fabric, making sure it is centered and/or aligned.

Secure the template to the fabric by taping around three edges only. Leave the fourth edge open.

3. Slip a piece of carbon paper, colored side down, through the opening so that it is sandwiched between the fabric and the design template. Tape down the fourth edge of the template.

4. Go over the design lines with a stylus or ballpoint pen, making sure to capture all the fine details. Work slowly and carefully; an accurately transferred design always results in a more beautifully stitched piece.

5. Lift one corner of the template and carbon paper to assure that the design has transferred successfully. Retrace any faint lines before removing the papers.

Pricking and Pouncing

Pricking and pouncing is the traditional method of transferring a design for embroidery. Appropriate for any type of fabric, this method is particularly effective on textured cloth, such as damask, shantung, and velveteen.

1. Trace or photocopy the embroidery design onto heavy vellum to create a template.

2. To make a pricker, insert the eye end of a #24 chenille needle into a cork.

3. Place the design template right side down on a piece of felt. Use the pricker to pierce holes along all design lines, spacing them about 1/16" (less than 2 mm) apart. Hold the pricker in an upright position to ensure clean perforations. Work slowly and carefully; an accurately transferred design always results in a more beautifully stitched piece.

4. Tape the fabric to a clean, flat work surface, or pin it to a padded ironing board. Place the perforated design template smooth side down on top of the fabric. Make sure the template is centered and/or properly aligned, and then secure it to the fabric with tape or pins.

5. Pour a small amount of pounce powder or grated colored chalk into a dish. Dip a cotton ball into the powder, and shake off any excess. Dab the powder onto the perforations. When the entire design has been covered, carefully remove the template.

6. To see the transferred outline more clearly, connect the dots using a water-soluble pen or tailor's chalk.

Removing Transfer Marks

The transfer marks produced by direct transfer, dressmaker's carbon, and pounce powder or grated chalk should be easily removable by hand washing, particularly if the fabric was pre-treated with spray starch. Use a mild detergent in cool water, and rinse well. To mark nonwashable fabrics, use pounce powder only and gently dust off the remaining residue with a soft brush when you are through.

Always test the chosen transfer method on a scrap of the project fabric and make sure the marks are removable before transferring the entire design. To run a test, cut a small swatch of the project fabric. Apply a test pattern—for example, your name—to the swatch using the selected transfer method. Test the washability by laundering in cool water with a mild detergent. If transfer marks remain on the fabric, select another method and repeat the testing. Continue the process until a desirable outcome is achieved.

TIPS AND TECHNIQUES

Every needle artist enjoys learning new tips to enhance the stitching process. The helpful hints gathered here will aid in your mastery and enjoyment of the art of embroidery.

⚜ Always work with short thread lengths, cut no longer than 20" (51 cm). Longer lengths are uncomfortable to manipulate and tend to fray or twist. Shorter lengths, not exceeding 12" (30 cm), are recommended for fibers that fray easily, such as silk ribbon.

⚜ Needle eyes are machine-cut, which creates a distinct front and back to the eye hole. If you experience difficulty threading a needle, try turning the needle around and threading it from the opposite side. Needle threaders can be helpful, too.

⚜ Cut—do not break—embroidery thread. Threads that are broken tend to fray and are difficult to thread into the eye of a needle. Make sure the end you insert into the needle eye is trimmed neatly.

⚜ Use a waste knot for tying on. Thread the needle, and make a knot at the end of the thread. Insert the needle into the fabric from the right side approximately 6" (15 cm) to the left of the starting point. Draw through so the knot rests on top of the fabric. Bring the needle up through the fabric at the selected starting point and begin stitching. When you are finished, clip the knot. On the underside, thread the tail into the needle and weave it into the back of the work.

⚜ To tie off, or end a thread, push the threaded needle through to the back of the fabric and weave it in and out of three or four neighboring stitches.

⚜ To mount very delicate fabrics such as organdy or swiss batiste in a round frame, use white tissue paper in addition to wrapping the inner hoop with cotton tape. Sandwich the fabric between single sheets of tissue, mount the paper and the fabric together, and then tear away the paper from the center of the hoop where the embroidery is to be done. The paper that remains acts as a protective barrier between the fabric and the frame.

⚜ Always remove the embroidered piece from a round frame at the end of each stitching session to prevent marking or stretching the fabric.

⚜ If a design falls too close to the edge or corner of the fabric, mounting the piece in an embroidery frame can seem impossible. The solution is to extend the fabric with a false edge. First, tear or cut a square of muslin that will extend at least 4" (10 cm) beyond the circumference of the frame. Next, center the edge or corner to be embroidered on top of the muslin. Baste the two layers of fabric together. Finally, carefully cut away the muslin behind the area to be embroidered. Mount the piece in the frame and stitch the design. Remove the basting stitches and muslin when the embroidery is completed.

⚜ Before embroidering a design that has been transferred by pricking and pouncing, spray the chalk marks with hairspray. Hairspray acts as a fixative to inhibit smudging and fading. A caveat: Never use hairspray on silk fabric, and spot-test all fabrics before attempting this method.

⚜ Always wash your hands and dry them thoroughly before each stitching session. Even if your hands appear clean, your natural skin oils and residues from hand lotions and cosmetics can cause soiling.

STITCH LIBRARY

Stem Stitch

STEP 1: Bring the thread to the front of the fabric on the left end of the design line (point A). Hold down the thread with your left thumb, and insert the needle into the fabric on the design line slightly to the right (point B). Bring the tip of the needle out midway between points A and B (point C).

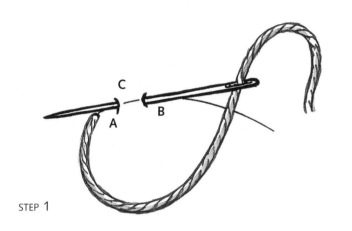

STEP 1

Continue holding down the thread with your thumb as you pull the thread through to set the first stitch.

STEP 2: Insert the needle into the fabric on the design line slightly to the right of point B. Bring the needle to the front again at point B (in exactly the same hole). Hold the thread down with your left thumb and pull the thread through to set the second stitch. Continue working the embroidery in this way. Try to make all of the stitches about ¹/8" (3 mm) in length.

To tie off, take the needle to the back at the end of the design line. Anchor the thread with three or four small loop knots.

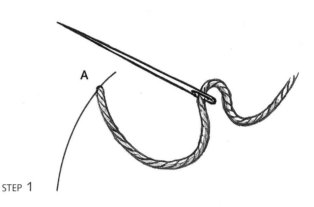

STEP 1

Backstitch

STEP 1: Bring the thread to the front of the fabric a short distance from the right end of the design line (point A).

STEP 2: Insert the needle into the fabric at the end of the design line (point B). Bring the tip of the needle out on the design line to the left of point A (point C). Pull the thread through to set the first stitch.

STEP 3: Reinsert the needle into the fabric at point A (in exactly the same hole). Bring the tip of the needle out on the design line to the left of point C (point D). Pull the thread through to set the second stitch. Continue working the embroidery in this way. Try to make all the stitches about ¹/16" (less than 2 mm) in length.

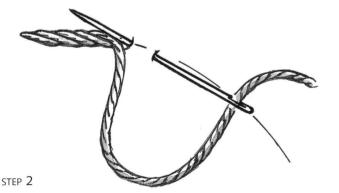

STEP 2

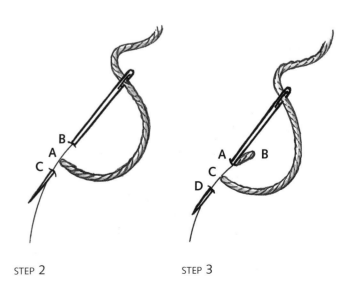

STEP 2   STEP 3

Interlaced Backstitch

Interlaced backstitch is actually a combination stitch. First, work the backstitch along the design line, and tie off.

Thread a #28 tapestry needle with the same or a contrasting thread. Bring the needle to the front of the fabric under the first stitch at the right end of the backstitched design line (between points A and B). Weave the needle in and out, alternately, under the backstitches. Take care not to pierce either the fabric or the backstitch threads.

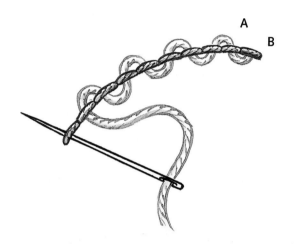

Straight Stitch

Bring the needle to the front of the fabric (point A). Insert the needle back into the fabric (point B) for the desired stitch length, and then bring it out at the beginning of the next stitch (point C). Straight stitches can be worked uniformly or irregularly, depending upon the effect you wish to achieve. It is best to keep them short in length and resting firmly against the ground fabric; they tend to snag when they are too long or too loose.

To tie off, take the needle to the back on the last stitch. Anchor the thread with three or four small loop knots.

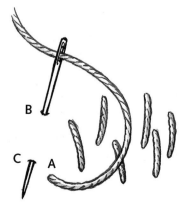

To tie off, take the needle to the back on the last stitch at the end of the design line. Anchor the thread with three or four small loop knots.

Double Backstitch (a.k.a. Shadow Work)

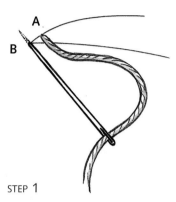

STEP 1

STEP 1: Bring the thread to the front of the fabric on the upper design line, a short distance from the tip of the shape (point

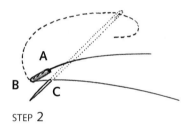

STEP 2

A). Insert the needle back into the fabric at the tip (point B). Pull through to complete the first stitch.
STEP   2:

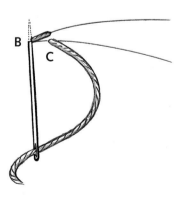

STEP 3

Reemerge on the lower line of the shape a short distance from the tip (point C).

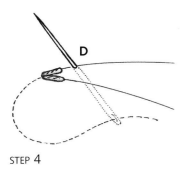

STEP 4

STEP 3: Reinsert the needle into the fabric at point B (in exactly the same hole). Pull the thread through to complete the second stitch.

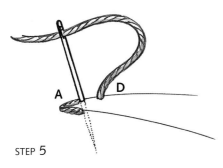

STEP 5

STEP 4: Bring the needle and thread to the front of the fabric

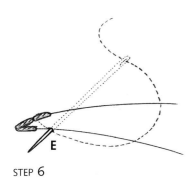

STEP 6

a short distance from point A (point D).

STEP 5: Reinsert the needle into the fabric at point A (in exactly the same

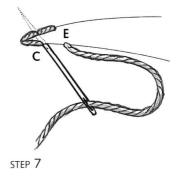

STEP 7

hole). Pull the thread through to complete the third stitch.

STEP 6: Reemerge on the lower line of the shape (point E).

STEP 7: Take the needle to the back at point C (in exactly the same hole). Pull the thread through to complete the fourth stitch.

STEP 8: Continue working the double backstitch in this man-

STEP 8

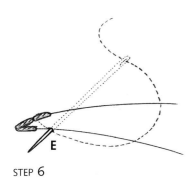

STEP 9

ner until you reach the end of the design lines. Try to make all the stitches about 1/16" (less than 2 mm) in length. In designs where one line is shorter, such as curved shapes, work the stitches on the shorter inside line slightly smaller so that each pair of backstitches moves in tandem.

STEP 9: On the reverse side, the threads should cross over one another. The crisscrossing threads create the "shadow" seen

Padded Satin Stitch/Satin Stitch

Follow all four steps for a padded satin stitch. For small shapes that don't require padding, omit step 2. For a plain satin stitch, omit steps 1 and 2 and simply embroider from edge to edge to fill the designated area.

STEP 1: Backstitch around the outline of the shape.

STEP 2: Fill the shape with tiny straight stitches that run perpendicular to the direction the satin stitch will be worked. This type of filling stitch is called "seeding."

STEP 3: Begin at the widest part of the shape. Use an up-and-down stabbing motion for the best results. Bring the needle to the front just outside the backstitched outline (point A). Pull the needle through, and take it to the back on the opposite side (point B), angling the needle under the outline. Continue working satin stitches very close together until half of the shape is covered.

STEP 4: Begin at the widest part again to work satin stitches over the remainder of the shape. The padded satin stitch is now complete.

To tie off, take the needle to the back and carefully weave the tail through the threads on the underside of the satin-stitched shape.

Overcast Stitch

Overcast stitch is a combination stitch that forms a smooth raised line. It is very useful for working monograms and the outlines of shapes.

First, work a row of stem stitch (page 15) along the design line. Then cover the row of stem stitch with tiny straight stitches. The straight stitches should be worked at a right angle to the padding row of stem stitches. A small amount of fabric should be picked up with each stitch that is taken, and the stitches should be made close together so that no ground fabric shows through.

To tie off, take the needle to the back and gently weave the tail through the threads on the underside of the overcast stitched line.

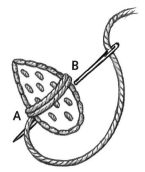

STEP 1

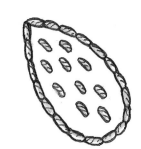

STEP 2

STEP 3

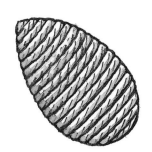

STEP 4

Detached Chain Stitch (a.k.a. Lazy Daisy)

STEP 1: Bring the needle to the front of the fabric at point A. Reinsert the needle into the fabric at point A (in exactly the same hole), and then bring the needle back up at point B.

STEP 2: Loop the thread counterclockwise so it passes under the needle. Push the needle through the fabric. Hold the thread loop lightly with your left thumb, and pull the needle gently away from you. Release the thread from your left thumb as the loop decreases in size.

STEP 3: Continue pulling the thread away from you until the loop lies flat against the surface of the fabric.

STEP 4: To anchor the stitch, take the needle to the back of the fabric just over the looped thread.

Chain Stitch

Chain stitch is actually a continuous row of detached chain stitches. Begin by working steps 1–3 for the detached chain stitch. (In other words, make a lazy daisy stitch, but don't anchor it down.)

STEP 5: Reinsert the needle into the fabric at point B (in exactly the same hole). Bring the needle back out at point C to finish the first stitch and to start the second stitch.

STEP 6: Continue working the embroidery in this way, making each stitch about 1/8" (3 mm) in length.

To tie off, insert the needle into the fabric just over the looped thread and take it to the back (Detached Chain Stitch, step 4). Anchor the thread with three or four small loop knots.

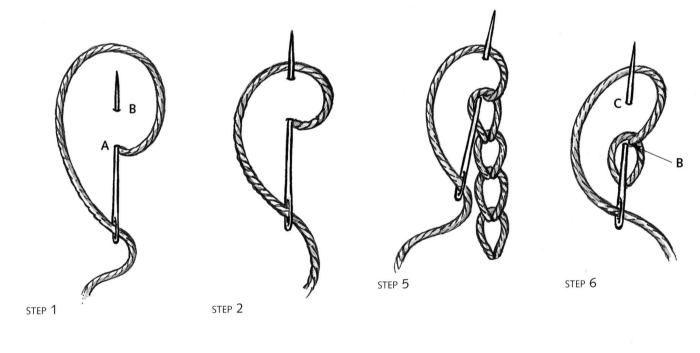

STEP 5   STEP 6

STEP 1   STEP 2

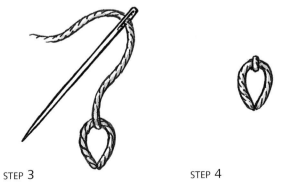

STEP 3   STEP 4

from the right side of the fabric.

To tie off, weave the tail of the thread through the crisscrossed stitches on the back, taking care not to pierce the fabric. Weaving back and forth three or four times will enhance the shadow effect.

Buttonhole Stitch

STEP 1: Bring the needle to the front of the fabric. Holding the thread down with your left thumb, insert the needle into the fabric at point A and come back out at point B. Still holding the thread down with your left thumb, pull the needle through the fabric and over the working thread.

STEP 2: Repeat the step 1 motion. The stitches in the illustra-

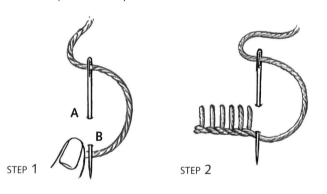

STEP 1   STEP 2

tion are slightly separated to clarify the technique, but you should work yours close together so that no ground fabric shows through.

To tie off, take the needle to the back on the last stitch at the end of the design line. Anchor the thread with three or four small loop knots.

Fly Stitch

STEP 1: Bring the thread to the front at point A. Insert the needle back into the fabric to the right at point B, and bring it through below at point C. Loop the thread counterclock-

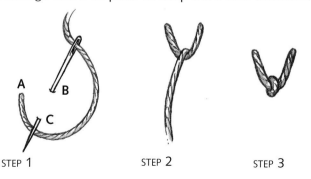

STEP 1   STEP 2   STEP 3

French Knot

STEP 1: Bring the needle to the front of the fabric at the place where the knot is to be positioned. Hold the thread taut between your left thumb and index finger approximately 1" (3 cm) away from the surface of the fabric.

STEP 2: Using your left hand, wrap the thread once around the needle.

STEP 3: Hold the thread taut again, and insert the point of the needle into the fabric one or two threads away from the starting point. Push the needle to the back of the fabric, all the while holding the thread down with your left thumb. Release your thumb as you pull the thread through to the back to set the French knot.

STEP 4: A completed French knot. If yours resembles a "Granny's bun" hairstyle, then you've done it right! For a more prominent knot, wrap the thread twice, or even three times, in step 2.

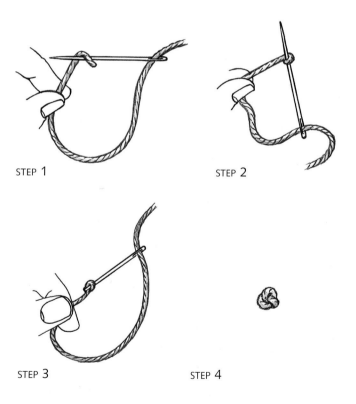

STEP 1   STEP 2

STEP 3   STEP 4

Bullion Stitch

STEP 1: Bring the needle to the front of the fabric at point A and pull the thread through. Insert the needle into the fabric at point B and bring it out again at point A, taking care not to split the thread. The thread should be to the right of the needle. Note: The distance between point A and point B should be the desired length of the finished bullion stitch.

STEP 2: From the underside, apply light pressure with the middle finger of your left hand to raise the point of the needle up off the fabric surface. Wrap the thread clockwise around the needle once, pulling tightly so the thread rests firmly against the surface of the fabric. Wrap the thread four or five more times around the shaft of the needle, keeping the wraps as even as possible.

STEP 3: Pull the thread with your right hand until the wraps are compacted together and rest firmly against the surface of the fabric. The wraps should be fairly snug, but not so tight as to bind the needle.

STEP 4: Hold the wraps in this position by placing your left thumb on top of them and your left index finger directly underneath on the underside of the fabric. Draw the needle and thread away from you, through the fabric and the wraps. The wraps should remain stationary as the thread slides through. Remove your left thumb and pull the needle and thread away from you again to ensure that the wraps are packed tightly together at the end of the thread emerging at point A.

STEP 5: Next, pull the thread firmly toward you. Use your left thumb to push the wraps snugly against the surface of the fabric.

STEP 6: The bullion stitch is now in place between points A and B. Anchor the stitch by inserting the needle into the fabric at point B and taking it to the back.

STEP 7: The finished bullion stitch looks like this. Note that the number of wraps required is determined by length of the stitch and the type of thread.

Tie off on the underside with three or four small loop knots.

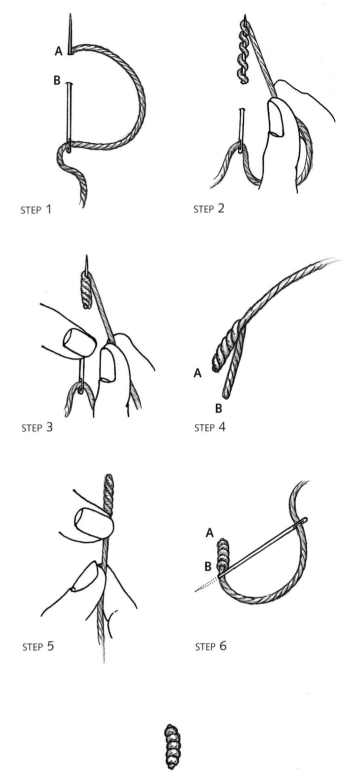

STEP 1

STEP 2

STEP 3

STEP 4

STEP 5

STEP 6

STEP 7

SECTION TWO: *Projects*

A snuggly wool throw for the living room, a charming tea cozy for the kitchen, elegantly monogrammed bed linens for the boudoir, fluffy cotton towels for the bath—these projects and twenty-six more (not to mention a host of variations) can be found on the following pages. The featured projects are organized room by room to give you lots of innovative ideas for decorating your own home. Venture with us into the embroidered living room, the embroidered kitchen and dining room, the embroidered bedroom and bath to see the delightful wonders you can create with needle and thread. A gallery section follows, brimming with accessories, small accent pieces, and gift ideas, many of which can be stitched in just a few hours.

To ensure your stitching satisfaction, each featured project has been carefully designed and tested, and then assigned a skill level:

Beginner, or Brush-up

Intermediate, or Some experience

Advanced, or Been stitching for quite a while

So that you can see at a glance what a project entails, we've also included a complete list of materials, a roster of the stitches used, a thread color guide, and a time estimate for completing the project. Photographs, clear step-by-step instructions, and full-color illustrations all work together to provide you with a foolproof strategy for successful stitching. You'll find additional stichery guidlines in the gallery section. Let's get started!

THE EMBROIDERED *Living Room*

The living room is a gathering place—a place where family members recount their daily experiences, as well as celebrate special occasions. Nothing is more inviting than a living room decorated to convey the unique personality of the family who dwells there. Whether you prefer furnishings that are boldly modern or elegantly traditional, you can personalize the main living area of your home with your own hand embroidery.

Imagine, if you will, a sleek, stark, contemporary room. The leather sofa by the fireplace looks inviting enough, but add a beautifully hand-stitched sage green blanket to the picture and you've created a homey spot that invites snuggling up with a good book. It's this kind of special attention to style and comfort that makes a house a home.

The projects in this chapter are designed to help you turn your living room into a haven for your family. Three of the projects—a silk dupioni cushion cover, linen curtain panels, and a wool throw—are stitched on ready-made pieces, making them all the more appealing because there's no finishing work involved. Simply embroider and enjoy! The fourth project—an exquisite mahogany storage box—requires only minimal finishing.

Of course, you can adapt any of the projects to fit your own décor. Changing the color palette or using a different stitch can significantly alter the appearance of a piece. Take a look at the project variations for ideas on working these designs into your overall decorating theme.

- GREEN WOOL THROW; OURS MEASURES 66" × 96" (168 CM × 244 CM)
- TWO 30-YARD (27-METER) SKEINS OF 3-PLY TWISTED SILK/MERINO WOOL YARN; WE USED SILK & IVORY 02 WHITE
- #24 CHENILLE NEEDLES
- EMBROIDERY SCISSORS
- 5" (13 CM) ROUND EMBROIDERY HOOP
- TAPE MEASURE
- SILK PINS (STRAIGHT PINS WITH GLASS OR METAL HEADS)
- PAPER SCISSORS
- TRANSPARENT TEMPLATE PLASTIC
- BLACK FINE-LINE FELT-TIP PEN
- TAILOR'S CHALK PENCIL

STITCH USED
STEM STITCH (PAGE 15)

 SKILL LEVEL
BEGINNER

 ESTIMATED TIME REQUIRED
8–10 HOURS

WHITE SILK/WOOL 3-PLY YARN

Wool Throw with leaves

Create a snuggly conversation piece for your living room when you embellish a ready-made throw with your hand embroidery. Single ivory-colored leaves drift downward against a ground of heathered green washable wool. The entire design is worked in stem stitch, making this a perfect project for a beginner. If the soft, natural palette isn't right for your living room's color scheme, try a bolder combination, like red and white or blue and yellow.

Wool Throw with leaves

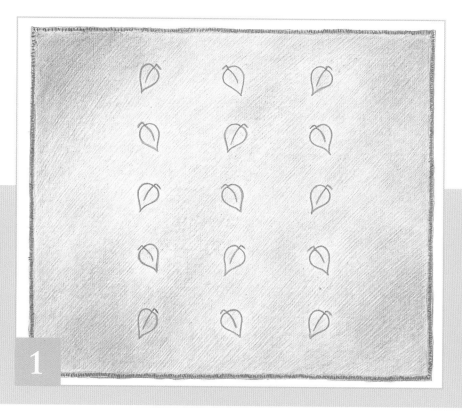

STEP 1 Using a black fine-line felt-tip pen, trace leaves A and B (page 112) three times each onto transparent template plastic. Cut out the six transparent templates with paper scissors. Fold the wool throw into quarters to locate the center, and mark with a straight pin. From the center, measure toward the longer edges of the throw 22" (56 cm) in each direction, and mark with pins. Pin two A and three B leaf templates—evenly spaced—along this 44" (112 cm) axis. Trace around the edge of each template with a tailor's chalk pencil. Remove the pins and templates. From the center axis, measure toward the shorter edges of the throw 20" (51 cm) in each direction, and mark with pins. Pin three A and two B leaf templates along each of these columns, even with the center axis leaves. Trace the leaf outlines. Remove all the pins and templates.

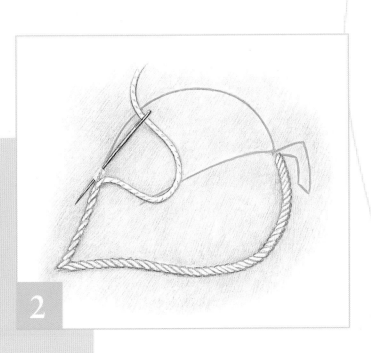

2

VARIATION

For an appealing twist, reverse the color scheme—stitch the leaves in sage green yarn on a winter white wool throw. Just for fun, make two companion throws, reversing their colors.

STEP 2   Embroider the fifteen leaves in stem stitch, using one strand of silk/wool yarn in a #24 chenille needle. Start each leaf at the base of the stem. Embroider around the perimeter, including the stem outline (or embroider a single-line stem, as shown in the project photo). End by stitching the vein down the middle of the leaf.

Use the template patterns on page 112.

- 16 × 16" (41 cm × 41 cm) SILK DUPIONI CUSHION COVER (PURCHASED OR SEWN FROM A COMMERCIAL PATTERN)

- SIX-STRAND COTTON EMBROIDERY FLOSS; WE USED ONE SKEIN EACH DMC #550, #552, #554, #3362, #3364

- #9 CREWEL NEEDLES

- EMBROIDERY SCISSORS

- 3" × 5" (8 cm × 13 cm) OVAL EMBROIDERY HOOP

- SILK PINS (STRAIGHT PINS WITH GLASS OR METAL HEADS)

- DRESSMAKER'S CARBON PAPER

- STYLUS

STITCH USED
CHAIN STITCH (PAGE 19)

SKILL LEVEL
BEGINNER

ESTIMATED TIME REQUIRED
10–12 HOURS

Silk Sofa Pillow

Dress up your sofa with this simple, elegant cushion. Regal purple flowers, sprigs of lavender berries, and lush green tendrils are joined by delicate violet swirls, all on a ground of butter-colored silk dupioni. The entire design is worked in chain stitch on a purchased cushion cover, making this project ideal for a beginning stitcher. The experienced seamstress may prefer to buy fabric yardage and custom-sew the pillow using a commercial pattern.

■ PURPLE DMC #550

■ VIOLET DMC #552

■ LAVENDER DMC #554

■ JADE DMC #3362

■ MOSS DMC #3364

Silk Sofa Pillow

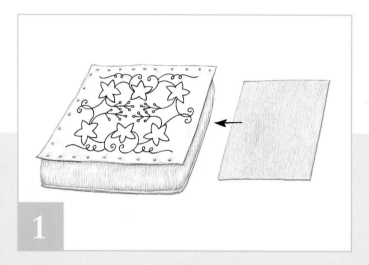

STEP 1 Unzip the cushion cover and remove the down pillow insert. Make two mirror-image copies of the template pattern (page 112). Tape both halves together, berries toward the middle, to fit the top of the cushion cover. Follow the Dressmaker's Carbon Transfer instructions (page 12) to mark the design on the front of the cushion cover.

STEP 2 Thread a #9 crewel needle with two strands of purple floss. Anchor the top layer of cushion fabric in the hoop so one of the star-shaped flowers is showing. Begin the chain stitch embroidery at the tip of one of the petals. When you reach the base of the petal, anchor the stitch to keep the inside angle sharp. Continue embroidering up the base of the adjacent petal, this time anchoring the stitch at the petal's tip. Continue embroidering all around in this manner until you reach the starting point. Tie off.

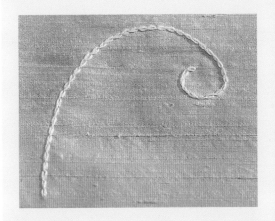

VARIATION

If your décor calls for a more neutral color palette, if you prefer a more understated look, or if you just want to simplify the project, try stitching the entire design in ecru-colored thread on taupe silk dupioni. You can also change the purple-and-green color scheme to match your existing home furnishings. Either way, you'll create a beautiful cushion.

DMC ECRU

STEP 3 Repeat step 2 to embroider each of the remaining star-shaped flowers. Embroider the other design lines in chain stitch in the same way, using violet floss for the connecting swirls, the two greens (one strand of each) for the tendrils, and lavender for the sprigs of berries.

Use the template pattern on page 112.

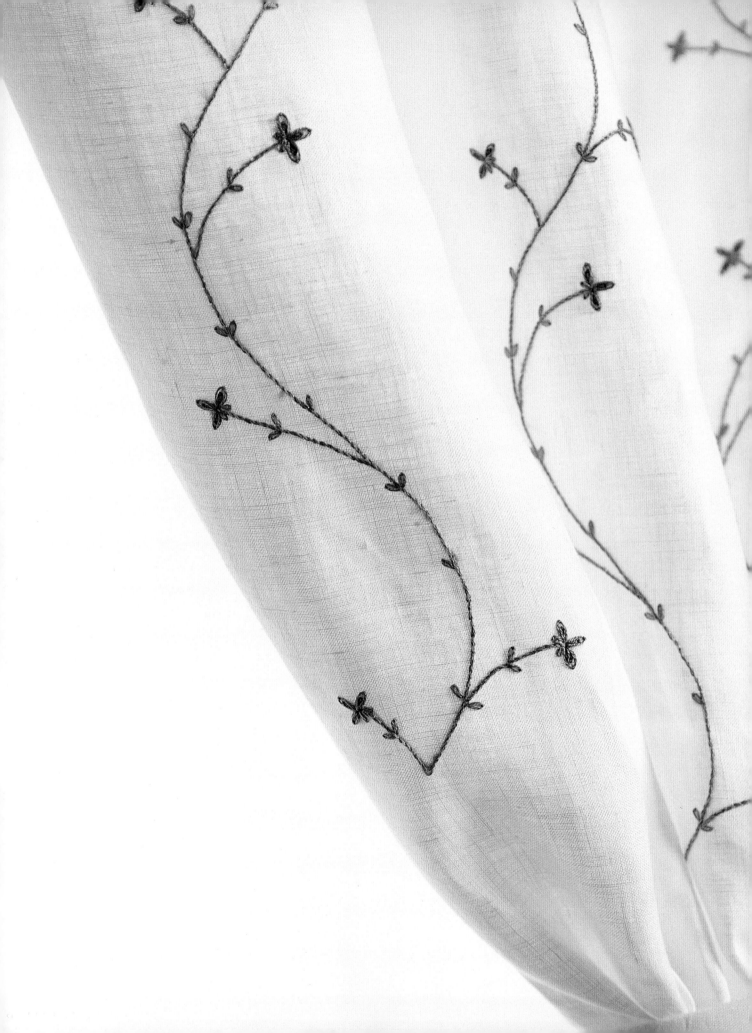

- WHITE LINEN TIE-ON CURTAIN PANELS, 54" × 96" (137 CM × 244 CM) (PURCHASED OR SEWN FROM A COMMERCIAL PATTERN)
- SIX-STRAND COTTON EMBROIDERY FLOSS; WE USED TWO SKEINS EACH OF DMC #320 AND #368, ONE SKEIN EACH OF #792 AND #793
- #10 CREWEL NEEDLES
- EMBROIDERY SCISSORS
- 5" × 7" (13 CM × 18 CM) OVAL EMBROIDERY HOOP
- SILK PINS (STRAIGHT PINS WITH GLASS OR METAL HEADS)
- DRESSMAKER'S CARBON PAPER
- STYLUS

STITCHES USED
STEM STITCH (PAGE 15)
DETACHED CHAIN STITCH (A.K.A. LAZY DAISY) (PAGE 19)

SKILL LEVEL
INTERMEDIATE

ESTIMATED TIME REQUIRED
APPROXIMATELY 45 MINUTES FOR EACH PATTERN REPEAT

Linen Curtain with Vine and Simple Flowers

Vivid periwinkle flowers and graceful green vines on a ground of white linen conjure up images of a South Seas paradise. The delicate embroidery on these airy, lightweight panels is sure to add a touch of mystery and tropical romance to your home. The design requires only two stitches: stem stitch and detached chain stitch. As always, you can alter the color scheme to suit the décor of your home, no matter where you live.

 DARK LEAF GREEN DMC #320

LEAF GREEN DMC #368

INDIGO DMC #792

 PERIWINKLE DMC #793

Linen Curtain with Vine and Simple Flowers

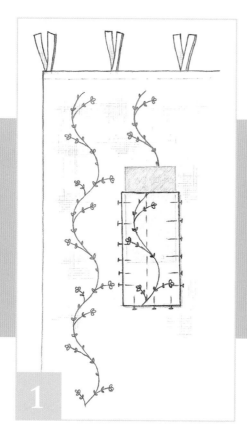

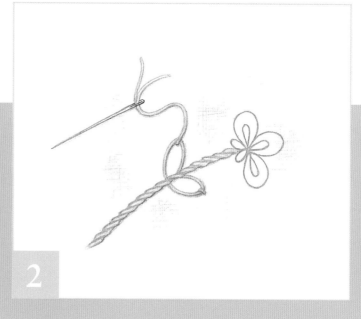

STEP 1 To mark this repeating design, begin at the upper left corner of the curtain panel. Following the instructions for Dressmaker's Carbon Transfer (page 12), mark one vine repeat (page 113) 1" (3 cm) below the top edge and centered between two pairs of tie tabs. Be sure to align the vertical dashed lines on the template with the fabric grain. Mark the second repeat directly below it so the vine continues uninterrupted. Continue in this way until the design has been transferred down the entire length of the panel; then repeat across the width.

STEP 2 For the foliage, thread a #10 crewel needle with the two green flosses (one strand of each). Starting at the top and working down, embroider the vines and stems in stem stitch. Work the leaves in detached chain stitch. Continue in this way until you have embroidered all of the foliage.

TIEBACKS

Tiebacks create subtle elegance, gracefully drawing a curtain aside to reveal an outdoor view. In this design, shadow work leaves sprout from a meandering stem stitch vine.

For each tieback, you will need two 9" × 18" (23 cm × 46 cm) pieces of white linen. Photocopy the template pattern (page 113) twice, once in mirror image, and tape the two halves together. Using the Dressmaker's Carbon Transfer method (page 12), mark the full tieback outline on one piece of linen, and mark the embroidery design on the half that will show at the window. Thread a #10 crewel needle with two strands of leaf green floss. First, embroider all of the vines in stem stitch. Next, work double backstitch to create the shadow work leaves. End by backstitching a vein down the center of each leaf.

To assemble the tieback, pin a 9" × 18" (23 cm × 46 cm) piece of medium-weight nonfusible interfacing to the back of the embroidered linen. Machine-baste along the marked outline. Place the embroidered linen and the plain linen pieces right sides together, and pin through all the layers. Machine-stitch along the basting line, leaving a 3" (8 cm) opening for turning. Trim the seam to 1/4" (5 mm) and zigzag to prevent fraying. Turn the tieback right side out, press the seams with a warm iron, and slip-stitch the opening closed. Tack small rings to each end. Your tieback is ready to use!

STEP 3 For the flowers, thread a #10 crewel needle with two strands of indigo floss. Make a small detached chain stitch in the center of each flower petal. Then, using two strands of periwinkle floss, make a larger detached chain stitch around each dark blue center.

Use the vine repeat template pattern on page 113.

 LEAF GREEN DMC #368

Storage Box with Grapes

Plump vineyard grapes dangle from a sturdy stem, enticing you to peek inside this handcrafted mahogany box to see what treasures you may find. The embroidery is worked on a ground of ivory silk dupioni. The look is luxurious, despite the fact that only three stitches are used to complete the design. The extensive use of satin stitch provides the ideal opportunity for perfecting your technique. After all, practice makes perfect!

MATERIALS

- MAHOGANY STORAGE BOX
- 10" × 10" (25 CM × 25 CM) SQUARE OF IVORY SILK DUPIONI
- SIX-STRAND COTTON EMBROIDERY FLOSS; WE USED TWO SKEINS OF DMC #550 AND ONE SKEIN EACH OF #869, #420, #3362, #3364
- #8 CREWEL NEEDLES
- #9 CREWEL NEEDLES
- EMBROIDERY SCISSORS
- 5" (13 CM) ROUND EMBROIDERY HOOP
- SILK PINS (STRAIGHT PINS WITH GLASS OR METAL HEADS)
- DRESSMAKER'S CARBON PAPER
- STYLUS
- IRON AND IRONING BOARD

STITCH USED
STEM STITCH (PAGE 15)
BACKSTITCH (PAGE 15)
SATIN STITCH (PAGE 18)

SKILL LEVEL
ADVANCED

ESTIMATED TIME REQUIRED
12–14 HOURS

PURPLE DMC #550

BROWN DMC #869

CHESTNUT DMC #420

JADE DMC #3362

MOSS DMC #3364

Storage Box with Grapes

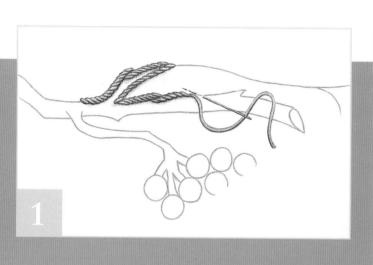

STEP 1 Iron the silk dupioni to remove any wrinkles. Follow the Dressmaker's Carbon Transfer instructions (page 12) to center and mark the grapes design (page 113) on the silk fabric. Thread a #9 crewel needle with one strand each of brown and chestnut floss. Secure the fabric in the embroidery hoop so that the woody vine is showing. Embroider the vine in stem stitch. Use the same brown threads to embroider the curly tendril to the right of the grapes in stem stitch.

STEP 2 For the leaves, thread a #9 crewel needle with two strands of jade floss. Work the outer portion of each leaf in satin stitch: bring the needle to the front of the fabric on the inner line, reinsert it on the outer line, and pull through. Continue stitching in this manner all around. Thread a #9 needle with one strand each of jade and moss floss. Embroider the leaf veins and the curled tendrils in stem stitch.

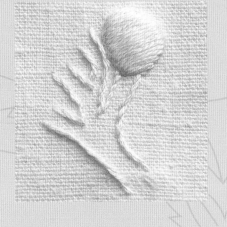

VARIATION

Use the grape design to create an heirloom similar to the antique tablecloth on the front cover of the book. Simply stitch the design in white thread on a white tablecloth, and voila! You'll have a keepsake that would make a welcome addition to any family's linen closet.

STEP 3 To outline the grapes, thread a #9 crewel needle with two strands of purple floss. Starting at the top of the cluster, carefully backstitch around each grape. Make the stitches 1/16" (less than 2 mm) in length to ensure a smooth edge for the satin stitch that will follow. To fill the grapes, thread a #8 crewel needle with three strands of purple floss. Once again, start at the top of the cluster and work downward. Bring the needle and thread out of the fabric at the lower left edge of the grape, reinsert it at the opposite edge, and pull the thread through. Note how the first stitch is angled across the widest part of the area to be filled. Fill the top half of the grape first, then the bottom half. Repeat to fill all of the grapes in the cluster.

Use template pattern on page 113.

THE EMBROIDERED *Dining Room and Kitchen*

What do you enjoy most about your kitchen? Is it baking chocolate chip cookies (with the help of the children) on a Saturday morning? Sitting down to a cup of coffee and the newspaper in the breakfast nook? Or is your kitchen primarily a work space where you prepare meals that are served in the dining room?

Kitchens and dining rooms are traditionally warm, inviting places where family members and friends share everything from after-school snacks to holiday dinners. Families are turning to the kitchen as their prime gathering place, and in many regions, the dining room has been incorporated into the kitchen's floor plan to create a large, casual area for entertaining. Whether your house features this newer building trend or you have a more traditional layout with a separate dining room, you can personalize the heart of your home with accessories you embroider by hand.

The four creative projects in this chapter are designed to add a big helping of charm to the culinary center of the home. Three of the projects—a tea cozy, kitchen towel, and tablecloth and napkins set—are stitched on ready-made pieces, making them all the more appealing because there's no finishing work involved. Simply embroider and begin using your new accessories! A fourth project—a charming patchwork chair cushion—requires some sewing and finishing.

You can adapt any of our projects to fit your décor simply by changing the color palette or using a different stitch. Because even small changes can significantly alter the appearance of a piece, take your time when evaluating your options. The suggested variation for each project can help you visualize the possibilities.

- WHITE LINEN TEA COZY (PURCHASED OR SEWN FROM A COMMERCIAL PATTERN)
- SIX-STRAND COTTON EMBROIDERY FLOSS; WE USED ONE SKEIN EACH OF DMC #550, #552, #727, #973, # 972, #905, #906
- #9 CREWEL NEEDLES
- EMBROIDERY SCISSORS
- 6" (15 CM) ROUND EMBROIDERY HOOP
- SILK PINS (STRAIGHT PINS WITH GLASS OR METAL HEADS)
- DRESSMAKER'S CARBON PAPER
- STYLUS

STITCH USED
STEM STITCH (PAGE 15)

SKILL LEVEL
BEGINNER

ESTIMATED TIME REQUIRED
2–3 HOURS

Pansy Tea Cozy

Add a touch of charm to your afternoon tea with this cheerful purple and yellow pansy. The entire design is worked in stem stitch on a purchased tea cozy, for a terrific beginner's project. The experienced seamstress has the option of buying white linen fabric yardage and sewing the cozy from a commercial pattern.

PURPLE DMC #550

VIOLET DMC #552

BUTTERCUP DMC #727

LEMON SUNSHINE #973

SUNFLOWER DMC #972

GRASS DMC #905

KEY LIME DMC #906

Pansy Tea Cozy

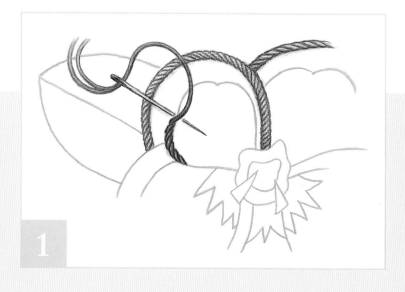

1

STEP 1 Remove the padded liner from the tea cozy. Follow the Dressmaker's Carbon Transfer instructions (page 12) to mark the pansy design on the front of the cozy. Thread a #9 crewel needle with two strands of violet floss. Embroider the outside line of each top petal in stem stitch. Change to two strands of purple floss to embroider the inner scalloped lines. For sharper definition between the scallops, pull the needle through to the back at the end of a scallop, reemerge one or two threads below and one or two threads to the left, and resume stitching. Work the three yellow petals the same way, using one strand each of buttercup and lemon sunshine floss for the outside line and two strands of buttercup for the inside line.

2

VARIATION

If you prefer pansies that are all one color, or if you're just looking for a way to simplify the project, try stitching the entire flower in purple. Another palette option is to let your favorite variety of pansy inspire your choice of thread colors. Either way, this cheerful addition to teatime is sure to make you smile!

STEP 2 Thread a #9 crewel needle with two strands of purple floss. Embroider the pansy's center "rings" and zigzags in stem stitch. For sharply defined zigzag points, repeat the step 1 technique: bring the needle to the back at the point, reemerge a few threads away, and resume stitching. Continue embroidering in this manner until all of the zigzag lines in the center of the pansy are stitched. Use two strands of sunflower floss to work the two triangular pistils in stem stitch. For the leaves, change to one strand each of grass and key lime floss. Outline each leaf in stem stitch.

Use the template pattern on page 114.

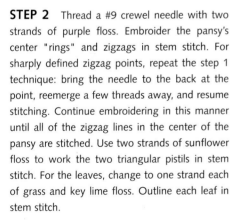

- COMMERCIAL PATTERN FOR A
  TIE-ON CHAIR CUSHION
- FOUR 6" × 6" (15 CM × 15 CM) SQUARES
  OF WHITE COTTON FABRIC
- COORDINATING FABRIC, TRIMS, ETC.
  (AS SPECIFIED ON PATTERN)
- SIX-STRAND COTTON EMBROIDERY FLOSS;
  WE USED ONE SKEIN OF DMC #321
- #9 CREWEL NEEDLES
- EMBROIDERY SCISSORS
- 4" (10 CM) ROUND EMBROIDERY HOOP
- SILK PINS (STRAIGHT PINS WITH GLASS
  OR METAL HEADS)
- RULER
- #2 PENCIL
- REMOVABLE TAPE
- SPRAY STARCH
- IRON AND IRONING BOARD
- DRESSMAKER'S SHEARS
- SEWING MACHINE

STITCHES USED
STEM STITCH (PAGE 15)
STRAIGHT STITCH (PAGE 16)

 SKILL LEVEL
BEGINNER

ESTIMATED TIME REQUIRED
1–2 HOURS FOR EACH MOTIF

 CHERRY RED DMC #321

Tie-on Chair Cushion with Cherries

Let clusters of bright red cherries lend their country charm to your kitchen.

This redwork project is perfect for beginners, with its easy-to-learn stitches:

stem stitch and straight stitch. To make the chair cushion, start with a

purchased pattern and adapt it for embroidery. You might try our checkerboard

arrangement, described in the instructions, or perhaps you'll position a cluster

of cherries in each corner or at the center. Some previous sewing experience is

advised. If sewing isn't your forte, find an experienced seamstress with whom

you can barter—you embroider something for her, and she, in return, can

assemble and finish a set of cushions for you.

Tie-on Chair Cushion with Cherries

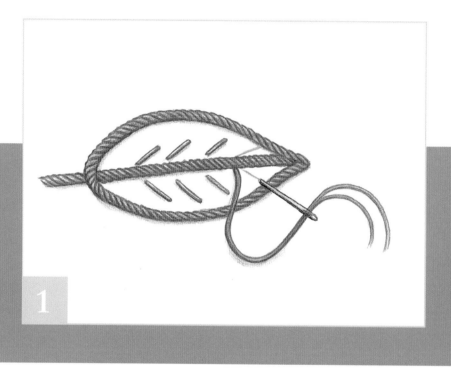

1

STEP 1 Follow the Direct Tracing Transfer instructions (page 12) to mark the cherry design (page 114) on each square of white fabric. Thread a #9 crewel needle with two strands of cherry red floss. Embroider each design, working the branch, stems, cherries, and, finally, leaves. Begin each leaf at the base, stitch all around, and then work the vein down the center. Embroider the tiny lateral veins in straight stitch, drawing the needle out at the center vein and reinserting it about 1/8" (3 mm) away. To avoid a thread shadow, slide the needle under the center vein stitching on the underside and bring it into position for the next lateral vein stitch.

VARIATION

The cherry design can be embroidered on dishtowels, café curtains, place mats, and napkins to create a complete kitchen ensemble. If you'd like more realistic cherries, simply add some color. Use lime peel green floss for the leaves and light brown floss for the branch and stems. Here's the complete palette:

CHERRY RED DMC #321

LIME PEEL DMC #3346

LIGHT BROWN DMC #434

STEP 2 Cut five 6" × 6" (15 cm × 15 cm) squares of coordinating fabric (we used a red-and-white check). Arrange the four embroidered squares and five coordinating squares in a checkerboard pattern. Pin the squares together in rows, and machine-stitch, using a ¹/2" (1 cm) seam allowance. Then join the rows together to create a 16" × 16" (41 cm × 41 cm) cushion top. Proceed with the purchased pattern and the additional supplies to complete the chair cushion.

Use the template pattern on page 114.

- LINEN DISH TOWEL
- SIX-STRAND COTTON EMBROIDERY FLOSS; WE USED ONE SKEIN EACH OF DMC #498, #321, #815, #434, #3346, #988
- #8 CREWEL NEEDLES
- #9 CREWEL NEEDLES
- EMBROIDERY SCISSORS
- 5" (13 CM) ROUND EMBROIDERY HOOP
- SILK PINS (STRAIGHT PINS WITH GLASS OR METAL HEADS)
- DRESSMAKER'S CARBON PAPER
- STYLUS
- IRON AND IRONING BOARD

STITCHES USED
CHAIN STITCH (PAGE 19)
BACKSTITCH (PAGE 15)
STEM STITCH (PAGE 15)
SATIN STITCH (PAGE 18)

SKILL LEVEL
INTERMEDIATE

ESTIMATED TIME REQUIRED
4 HOURS

Kitchen Towel with Apple Motif

Enjoy an orchard harvest year-round with this embellished kitchen towel.
Our juicy Red Delicious apple looks realistic enough to pluck off the window-pane-checked linen ground, but closer examination reveals a fruit that is not the eating kind. Four simple stitches are used: chain stitch, backstitch, stem stitch, and satin stitch. Quick and easy to embroider on a purchased linen towel, the apple will surely add a cheerful note to your kitchen. Your handwork would also make a thoughtful gift for a new bride.

■ DARK RED DMC #498

■ CHERRY RED DMC #321

■ CRIMSON DMC #815

■ LIGHT BROWN DMC #434

■ LIME PEEL DMC #3346

■ GREEN APPLE DMC #988

Kitchen Towel with Apple Motif

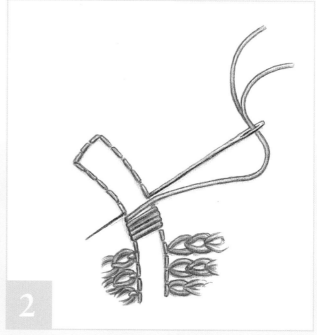

STEP 1 Fold the kitchen towel in half length-wise and crease lightly with a warm iron. Follow the Dressmaker's Carbon Transfer instructions (page 12) to mark the apple design (page 114) on the towel; center it on the crease about 7" (18 cm) above the bottom edge. Thread a #8 crewel needle with one strand each of dark red, cherry red, and crimson floss. Beginning at the stem, work chain stitch on the apple outline all around. Continue stitching concentric rings of chain stitch until the entire shape is filled. Tie off.

STEP 2 For the stem, thread a #9 crewel needle with two strands of light brown floss. Backstitch along the apple stem outline. To fill in the stem, use satin stitch. Bring the needle and thread to the front of the fabric along one edge, insert the needle into the fabric on the opposite edge, and pull the thread through to the back to complete the first stitch. Continue to work satin stitch over the top half of the stem, then start at the middle and work in the other direction to satin-stitch the bottom half. Tie off.

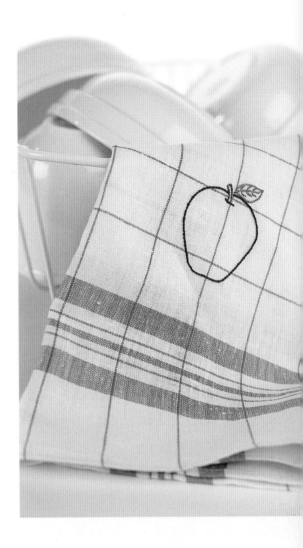

STEP 3 For the leaf, thread a #9 crewel needle with one strand each of lime peel green and green apple floss. Starting at the base, embroider the leaf outline in stem stitch. To fill in the leaf, use satin stitch. Bring the needle and thread to the front of the fabric on the vein, insert the needle back into the fabric on the leaf outline, and pull the thread through to the back to complete the first stitch. Note how the stitch lies at a slight angle. Continue working satin stitches in this manner until you reach the leaf tip, and tie off. Embroider the area on the opposite side of the vein in the same way, and then embroider the lower half of the leaf, working toward the base. Finally, embroider the leaf vein in stem stitch, using two strands of green apple floss. Tie off.

Use the template pattern on page 114.

VARIATION

For a quicker finish, work stem stitch on the design outlines only. If you prefer a different variety of apple, simply change the thread colors to suit your taste. Golden Delicious apple lovers can replace our vibrant reds with a softer yellow palette. Granny Smith aficionados could use vivid greens on a plain white towel for a striking effect.

Scrollwork Tablecloth and Napkins

MATERIALS

- WHITE LINEN TABLECLOTH AND NAPKINS
- SIX-STRAND COTTON EMBROIDERY FLOSS; WE USED THREE SKEINS OF DMC #841
- #9 CREWEL NEEDLES
- EMBROIDERY SCISSORS
- 5" (13 CM) ROUND EMBROIDERY HOOP
- SILK PINS (STRAIGHT PINS WITH GLASS OR METAL HEADS)
- TAPE MEASURE
- #2 PENCIL
- REMOVABLE TAPE
- SPRAY STARCH
- IRON AND IRONING BOARD

STITCH USED
DOUBLE BACKSTITCH (A.K.A. SHADOW WORK) (PAGE 16)

 SKILL LEVEL
INTERMEDIATE

 ESTIMATED TIME REQUIRED
2 HOURS FOR EACH TABLECLOTH MOTIF
45 MINUTES FOR EACH NAPKIN MOTIF

 TAUPE DMC #841

Dinner guests will feel like royalty when they sit down to a meal at your splendidly appointed table. Embroider these elegant designs for a stunning addition to your linen trousseau. In double backstitch, also known as "shadow work," the designs are worked on ready-made table linens, making this project even more desirable because there's no finishing required.

Tablecloth and Napkins

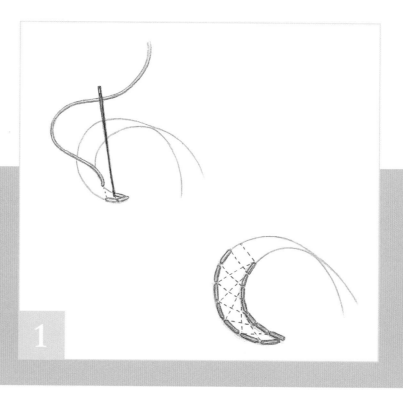

1

STEP 1 Apply a light coat of spray starch to the entire surface of the tablecloth. Allow two or three minutes for the fabric to absorb the starch, then iron the tablecloth dry. Follow the Direct Tracing Transfer instructions (page 12) to mark the large scroll design (page 115) on each corner of the tablecloth. Thread a #9 crewel needle with two strands taupe floss. Start in the lower left quadrant of the design. Embroider the scroll lines in double backstitch, remembering to work stitches on the inside curve smaller than those on the outside, in order to keep each pair of stitches moving along at the same pace. Continue until all four corner motifs are embroidered.

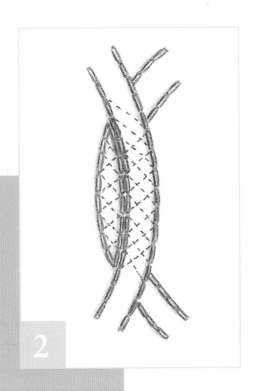

2

STEP 2 Starch and iron each napkin as you did for the tablecloth. Fold the napkin in half diagonally, finger-press lightly at one corner, and unfold. Mark the small scroll design (page 115) on the napkin corner, using the crease for alignment. Thread a #9 crewel needle with two strands of taupe floss. Embroider the scroll lines on the left half of the design in double backstitch. Tie off. Work the right half the same way. When you reach the intersection of the design lines, work one line of double backstitches directly alongside the existing stitches, so that they share the same holes. Work the remainder of the motif as usual. Tie off. Repeat for each napkin.

Use the template patterns on page 115.

VARIATION

To re-create the elegant look of antique linens, try embroidering scroll designs in padded satin stitch, shown above in progress. The embroidery will take considerably longer to complete than shadow work, but the finished linens will possess an unmistakable heirloom quality. Choose a thread color that coordinates with your fine china.

PASTEL BLUE DMC #800

THE EMBROIDERED *Bedroom*

The English translation of boudoir is "private retreat." Certainly, with today's busier-than-ever lifestyles, there is no better place for recharging from life-on-the-run than the privacy of your own bedroom. Transform yours into a relaxing, personal sanctuary—a place where you feel as comfortable working on your needlework projects as you do drifting off to sleep on your hand-embroidered bed linens.

Regardless of the decorative style of your bedchamber, the presence of your own hand embroidery will make it feel even more snug and intimate. Outside distractions will melt away when you cuddle up in a soft, warm, monogrammed blanket, or sink your head into a pillow embroidered with spring's most delicate blossoms. Seeing beautiful stitchery up close, every day, will rejuvenate your soul.

Three of the projects in this chapter—crisp bed linens, a cuddly Merino wool blanket, and a hemstitched cotton bedskirt—started out as ready-made pieces. It doesn't matter what size bedding you use as the designs are readily adaptable. Just choose the styles and colors that are appropriate for your furnishings and décor, and then take them one beautiful step further by adding touches of hand embroidery. The remaining project—a delicate silk lampshade—requires a little bit of finishing that you can easily do yourself. The results are both unique and exquisite.

If you're interested in trying out some different color options, or perhaps are seeking a project that requires less stitching time, be sure to consider the project variations.

- WHITE LINEN HEMSTITCHED SHEET AND PILLOWCASES
- SIX-STRAND COTTON EMBROIDERY FLOSS; WE USED TWO SKEINS OF DMC #597 AND ONE SKEIN EACH OF DMC #598 AND #3046
- #9 CREWEL NEEDLES
- #7 BETWEENS NEEDLES
- EMBROIDERY SCISSORS
- 5" × 7" (13 CM × 18 CM) OVAL EMBROIDERY HOOP
- SILK PINS (STRAIGHT PINS WITH GLASS OR METAL HEADS)
- FRAY PREVENTER MEDIUM
- RULER
- TAPE MEASURE
- #2 PENCIL
- TRACING PAPER
- BLACK FINE-LINE FELT-TIP PEN
- REMOVABLE TAPE
- SPRAY STARCH
- IRON AND IRONING BOARD

STITCHES USED
DOUBLE BACKSTITCH (PAGE 16)
FRENCH KNOT (PAGE 20)

SKILL LEVEL
INTERMEDIATE

ESTIMATED TIME REQUIRED
8 HOURS FOR TOP SHEET
4 HOURS FOR EACH PILLOWCASE

AQUAMARINE DMC #597

ICY AQUA DMC #598

WHEAT DMC #3046

Bed Linens

Dress your bed for dreaming with exquisitely hand-embroidered bed linens.

Icy aqua diamonds and bold aquamarine flowers form a simple repeat pattern

for edging a top sheet and pillowcases. Add an optional wheat-colored

monogram, pictured on page 65, using our custom alphabet. When you

start with ready-made hemstitched linens, all you do is add your handwork.

Sweet dreams!

Bed Linens

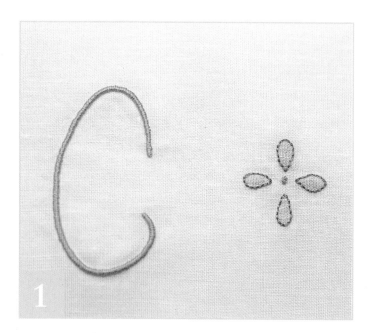

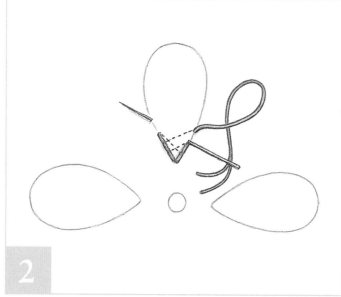

STEP 1 Apply a light coat of spray starch to the entire surface of the top sheet. Allow two or three minutes for the fabric to absorb the starch, then iron the sheet dry. Measure in 7 1/2" (19 cm) from the middle of the top edge, and mark with a pin. Follow the Direct Tracing Transfer instructions (page 12) to mark the flower-and-diamond design (page 115) across the sheet; center the middle flower at the pin. Thread a #9 crewel needle with one strand each of aquamarine and icy aqua floss. Working from left to right, embroider each diamond motif in double backstitch. Remember to work the stitches on the inside diamond smaller, in order to keep each pair of backstitches moving along at the same pace. Weave in the ends to enhance the shadow effect.

STEP 2 Thread a #9 crewel needle with two strands of aquamarine floss. Working from left to right, embroider the flowers in double backstitch. Work each petal individually, starting at the pointed end and tying off after the petal is completed. To enhance the shadow effect, repeat the step 1 technique for weaving in the ends.

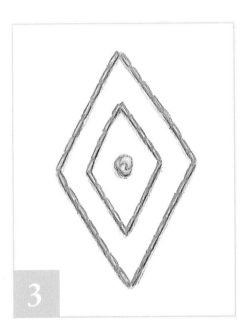

STEP 3 Thread a #7 betweens needle with two strands of wheat floss. Knot the thread 3" (8 cm) from the end. Working from left to right, embroider a 3-wrap French knot in the middle of each diamond and each flower motif. After drawing the thread to the back, clip it 3" (8 cm) from the end. Tie the two 3" (8 cm) tails together in a square knot, trim 1/4" (5 mm) from the knot, and seal the knot with a a dot of fray preventer.

Repeat steps 1–3 to mark and embroider each pillowcase.

Use the template pattern on page 115.

VARIATION

Make your bed linens even more sumptuous with an elegant embroidered monogram. To make the embroidery template, omit the center flower and trace one of the letters from our monogram alphabet (page 125) in that spot instead; you may wish to open up the spacing around the letter a little bit. Once you've made your template, transfer the design to your bed linens. Embroider the monogram in overcast stitch (page 18), using two strands of wheat floss.

- IVORY MERINO WOOL BLANKET; OURS MEASURES 66" × 96" (168 CM × 244 CM)
- BRODER MEDICIS WOOL THREAD; WE USED THREE SKEINS OF DMC #8685
- #24 CHENILLE NEEDLES
- EMBROIDERY SCISSORS
- 5" × 7" (13 CM × 18 CM) OVAL EMBROIDERY HOOP
- SILK PINS (STRAIGHT PINS WITH GLASS OR METAL HEADS)
- RULER
- #2 PENCIL
- TRACING PAPER
- BLACK FINE-LINE FELT-TIP PEN
- POUNCE POWDER (OR CHALK POWDER)
- COTTONBALLS
- FINE-LINE WATER-SOLUBLE MARKING PEN
- PADDED IRONING BOARD

STITCHES USED

STEM STITCH (PAGE 15)

PADDED SATIN STITCH (PAGE 18)

OVERCAST STITCH (PAGE 18)

 SKILL LEVEL

ADVANCED

ESTIMATED TIME REQUIRED

7–8 HOURS

 RASPBERRY DMC #8685

Monogrammed Blanket

Add warmth and elegance to your bedroom with a beautiful personalized blanket. Striking raspberry-colored plumes flank the bold monogram, all on a ground of ivory merino wool. Three easy stitches are used to create this sumptuous design: stem stitch, overcast stitch, and padded satin stitch. As always, the color scheme can be modified to suit the décor of your home.

Monogrammed Blanket

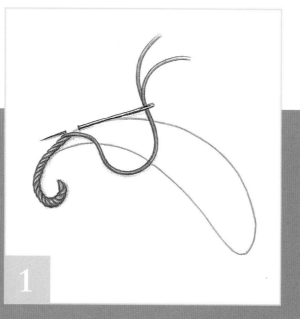

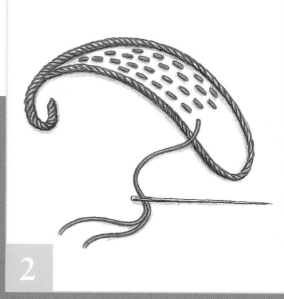

STEP 1 Locate the middle of the blanket's top edge, measure in 7" (18 cm) and mark with a pin. Follow the Pricking and Pouncing instructions (page 13) to transfer the open scroll design (page 116) to the blanket, at the pin marking. In addition, mark two closed scrolls evenly spaced on either side. Thread a #24 chenille needle with two strands of raspberry broder medicis wool. Working from left to right, outline each shape in stem stitch. To stitch the curved plumes, begin at the curled tail and work up the outside curve and around. Stitch each teardrop plume starting and ending at the pointed tip.

STEP 2 Thread a #24 chenille needle with two strands of raspberry broder medicis wool. Beginning at the pointed end of each plume, fill the interior with tiny, randomly spaced straight stitches not exceeding 1/8" (3 mm) in length. Tie off. This technique, called "seeding," provides a padding for the satin stitching that will follow. Seeding stitches are always worked perpendicular to the satin stitches that will cover them.

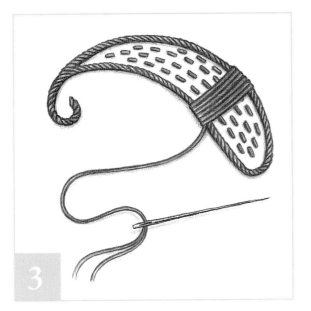

3

STEP 3 For the satin stitch, thread a #24 chenille needle with two strands of raspberry broder medicis wool. Starting in the middle of each plume, bring the needle and thread to the front of the fabric on the inside curve. Insert the needle into the fabric on the outside curve, and pull the thread through to the back to complete the first stitch. Continue working satin stitch over the seeding to cover the broader end of the plume. Start at the middle and work in the other direction to cover the narrower end.

4

STEP 4 To add a monogram, draft a 2 1/2" (6 cm) line on tracing paper. Place the paper on monogram alphabet (page 125), center the line horizontally on the desired letter, and trace. Follow the Pricking and Pouncing instructions (page 13) to transfer the monogram to the blanket, centering it between the two embroidered open scrolls. Thread a #24 chenille needle with two strands of raspberry broder medicis wool. Embroider the monogram in overcast stitch.

Use the template patterns on page 116 and the monogram alphabet on page 125.

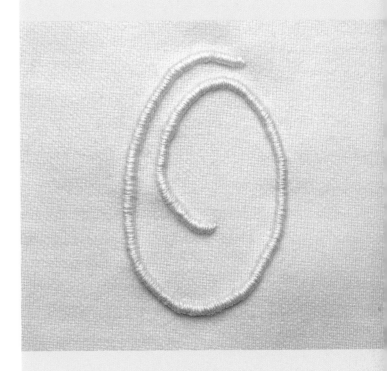

VARIATION

For an understated look, stitch the entire design in ivory-colored yarn. The white-on-white palette will focus attention on the embroidery's texture.

Cherry Blossoms Lamp Shade

MATERIALS

- SELF-ADHESIVE LAMP SHADE
- IVORY SILK DUPIONI FABRIC
  (TO COVER LAMP SHADE)
- IVORY-COLORED TRIM
  (FOR BOTTOM EDGE OF LAMP SHADE)
- IVORY-COLORED FLEXIBLE BINDING
- SIX-STRAND COTTON EMBROIDERY FLOSS;
  WE USED ONE SKEIN EACH OF DMC #3863,
  #818, #3326
- #8 CREWEL NEEDLES
- #9 CREWEL NEEDLES
- EMBROIDERY SCISSORS
- 5" (13 CM) ROUND EMBROIDERY HOOP
- TAPE MEASURE
- SILK PINS (STRAIGHT PINS WITH GLASS
  OR METAL HEADS)
- DRESSMAKER'S CARBON PAPER
- STYLUS
- IRON AND IRONING BOARD

STITCHES USED

STEM STITCH (PAGE 15)
BUTTONHOLE STITCH (PAGE 20)
DETACHED CHAIN STITCH
    (A.K.A. LAZY DAISY) (PAGE 19)
FRENCH KNOT (PAGE 20)

SKILL LEVEL
INTERMEDIATE

ESTIMATED TIME REQUIRED
2–3 HOURS

Embroider our delicate pink cherry blossoms on a ground of ivory silk dupioni and experience for yourself the calming effects of Asian-influenced design. This classic motif requires four stitches: stem stitch, buttonhole stitch, detached chain stitch and French knot. Work the embroidery for one of the self-adhesive lamp shades that are readily available in today's marketplace. If you have previous experience or expertise in lamp shade crafting, you might adapt the needlework to one of your own shade designs.

 COCOA DMC #3863

 COTTON CANDY DMC #818

 ROSE BLUSH DMC #3326

Cherry Blossoms Lamp Shade

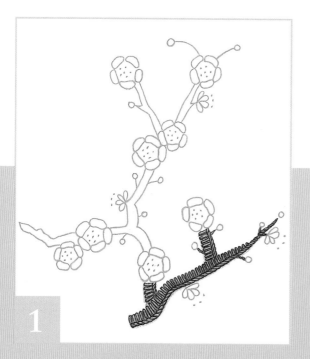

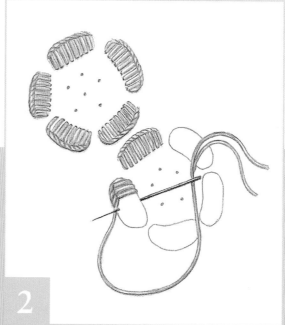

STEP 1 Iron the silk to remove any wrinkles. Mark the lamp shade outline on the silk, following the manufacturer's instructions. Mark the cherry blossom design (page 116) on the center front of the lamp shade, using the Dressmaker's Carbon Transfer method (page 12). Thread a #9 crewel needle with two strands of cocoa floss. Embroider the branches in buttonhole stitch. Add the small twigs in stem stitch.

STEP 2 For the cherry blossoms, thread a #9 crewel needle with two strands of cotton candy floss (for a more dramatic look, use two strands of rose blush floss). Work the A blossoms in buttonhole stitch, tying off after each flower. Work the B blossoms the same way, using one strand each of cotton candy and rose blush floss.

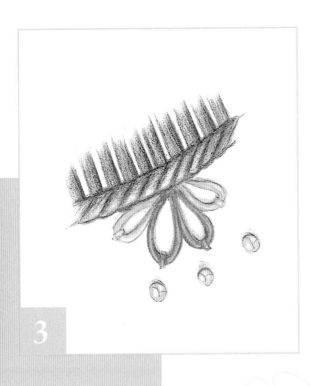

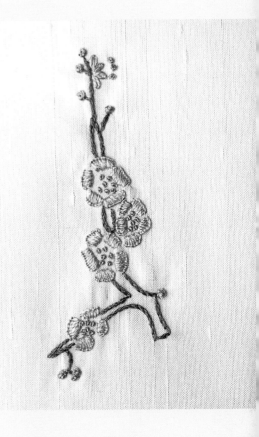

3

STEP 3 For the closed buds, thread a #8 crewel needle with three strands of rose blush floss. Work a 3-wrap French knot for each bud. For the slightly opened buds, thread a #9 crewel needle with two strands of rose blush floss. Work the two inner petals of each bud in detached chain stitch. Change to two strands of cotton candy floss to work the two outside petals. Work the three dots as three one-wrap French knots. Follow the manufacturer's instructions to assemble and trim the lamp shade.

Use the template pattern on page 116.

VARIATION

Embroider an abbreviated version of our classic Asian-inspired cherry blossoms around the bottom edge of a small lamp shade. For a lighter look, use stem stitch instead of buttonhole stitch for the branch.

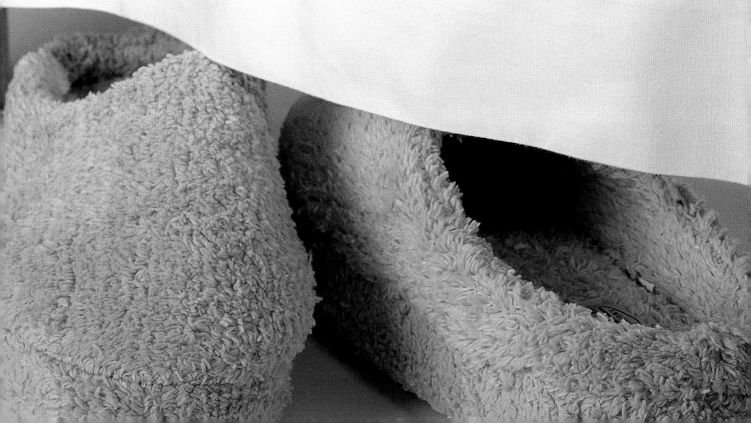

- WHITE COTTON HEMSTITCHED BED SKIRT
- SIX-STRAND COTTON EMBROIDERY FLOSS; WE USED ONE SKEIN EACH OF DMC #3348, #772, # 3753, #341, AND BLANC
- #9 CREWEL NEEDLES
- #7 BETWEENS NEEDLES
- EMBROIDERY SCISSORS
- 5" × 7" (13 CM × 18 CM) OVAL EMBROIDERY HOOP
- SILK PINS (STRAIGHT PINS WITH GLASS OR METAL HEADS)
- DRESSMAKER'S CARBON PAPER
- STYLUS

STITCHES USED

STEM STITCH (PAGE 15)

DETACHED CHAIN STITCH
    (A.K.A. LAZY DAISY) (PAGE 19)

FRENCH KNOT (PAGE 20)

 SKILL LEVEL
BEGINNER

 ESTIMATED TIME REQUIRED
2–3 HOURS

Floral Garlands Bedskirt

What better way to restore your spirit after a hectic day than by spending a quiet moment or two working on your latest embroidery project? Here's a design that will make your bedroom more restful and inviting. Add graceful floral garlands to a hemstitched bedskirt, and you quickly transform your bed into the focal point of the master suite. We selected peaceful shades of blue and green to play against the ground of whisper-soft white cotton—the perfect color scheme for fashioning your own personal retreat. Only three stitches are required: stem stitch, detached chain stitch, and French knot.

SPRING GREEN DMC #3348

SEA GREEN DMC #772

ARCTIC BLUE DMC #3753

CORNFLOWER DMC #341

WHITE DMC BLANC

Floral Garlands Bedskirt

STEP 1 Make two mirror-image copies of the bedskirt template pattern (page 117). Tape both halves together, with the six-petaled flower in the middle. Follow the Dressmaker's Carbon Transfer instructions (page 12) to mark the floral design around the perimeter of the bedskirt, 3" (8 cm) above the hemmed edge. We transferred a total of eight designs: three along each side of the bed and two at the foot of the bed. Thread a #9 crewel needle with one strand each of spring green and leaf green floss. Starting at the left end of each design, embroider the vine and stems in stem stitch. Work the leaves in detached chain stitch. Continue until all of the foliage has been completed.

STEP 2 For the five-petal flowers, thread a #9 crewel needle with one strand each of arctic blue and cornflower floss. Work each petal in detached chain stitch. Work each six-petal flower the same way, except use two strands of cornflower floss.

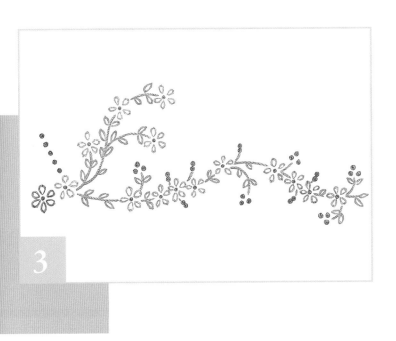

VARIATION

Short on time, but still love the look of hand-embroidered bed linens? Then stitch this abbreviated version of our floral garland motif. Using just one color of thread is another way to make the stitching go faster. We chose a vibrant yellow floss so that every day would begin with a little bit of sunshine!

BUTTERCUP DMC #727

STEP 3 For the French knots, thread a #7 betweens needle with two strands of white floss. Work a 2-wrap French knot at the center of every flower. Change to two strands of arctic blue floss. Work the flower sprigs with two buds as two 2-wrap French knots. Work the flower sprigs with three buds as three 3-wrap French knots. Change to two strands of cornflower floss. Work each of the nine dots at the center of the design as a 3-wrap French knot.

Use the template pattern on page 117.

THE EMBROIDERED *Bath*

Shiny brass fixtures that glow by candlelight, smooth ceramic tile that's cool to the touch, an antique claw foot tub filled to the brim with fragrant bubbles—here are the makings of a tranquil refuge where you can relax, pamper yourself, and restore vigor to both body and spirit. Today's bath is more than just a place to take a quick shower; it's a luxuriant retreat that lets you forget life's hassles.

Your own mini-spa will become even more inviting when you add hand-embroidered accessories and bath linens. Whether your taste runs to perky geometric stitchery on a bath towel or classic scroll embroidery on the back of a silver hand mirror, the opportunities for embellishing ordinary as well as specialty bath items are numerous.

Three of the projects in this chapter—fluttery café curtains, soft, plush bath towels, and a sassy shower curtain—are all stitched on ready-made pieces. As with so many of our projects, the appeal of embellishing a purchased item is undeniable. There's no further sewing or assembly, and you can begin using your new piece right away, as soon as the embroidery is finished. If you are more experienced in your stitching, you may want to try our fourth project, an elegant vanity set done in shadow work embroidery. It requires minimal finishing that you can easily do yourself.

There are many ways to adapt our bathroom designs. Different thread colors, different stitches, or even a change of background fabric can give a design a whole new look. Before you try something new, take a look at the project variations for some fresh takes on your spa renovation.

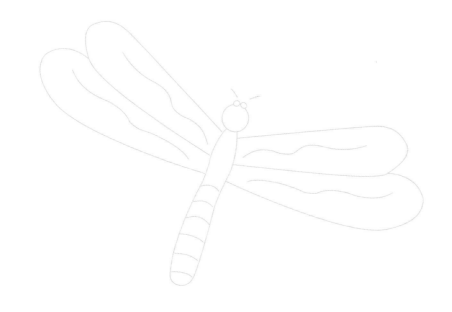

- WHITE COTTON CAFÉ CURTAIN PANELS; OURS MEASURE 42" × 30" (107 CM × 76 CM) EACH
- METALLIC BLENDING FILAMENT; WE USED ONE SPOOL EACH OF KREINIK METALLIC BLENDING FILAMENTS #023 AND #012
- SIX-STRAND COTTON EMBROIDERY FLOSS; WE USED ONE SKEIN EACH OF DMC #208, #209, #702, #703
- #8 CREWEL NEEDLES
- EMBROIDERY SCISSORS
- 5" (13 CM) ROUND EMBROIDERY HOOP
- TAPE MEASURE
- SILK PINS (STRAIGHT PINS WITH GLASS OR METAL HEADS)
- #2 PENCIL
- REMOVABLE TAPE
- SPRAY STARCH
- IRON AND IRONING BOARD

STITCHES USED
STEM STITCH (PAGE 15)
STRAIGHT STITCH (PAGE 16)
FRENCH KNOT (PAGE 20)

SKILL LEVEL
BEGINNER

ESTIMATED TIME REQUIRED
5–6 HOURS FOR EACH PANEL

café curtains

A glittering dragonfly hovers above a lush green meadow, beckoning you to come out and play. Spring will be forever in the air when you add a pair of these gauzy white cotton café curtains to your bath. You'll be amazed at how quick and easy they are to stitch! The design requires three basic stitches: stem stitch, straight stitch, and French knot. As an added bonus, the motif is embroidered on ready-made curtains so there's no finishing involved—just stitch your curtains, hang them in a sunny window, and enjoy!

LILAC METALLIC KREINIK #023

GRAPE METALLIC KREINIK #012

GRAPE DMC #208

LILAC DMC #209

PARROT GREEN DMC #702

GRASSHOPPER DMC #703

café curtains

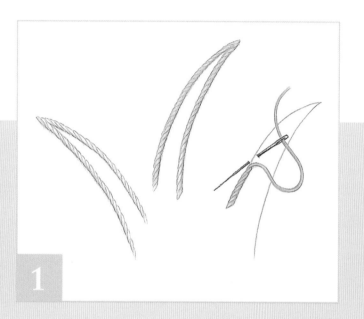

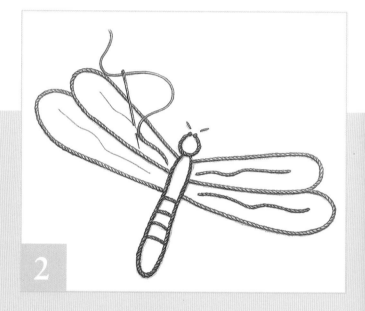

STEP 1   Apply a light coat of spray starch to the entire surface of one curtain panel. Allow two or three minutes for the fabric to absorb the starch, then iron the panel dry. Follow the Direct Tracing Transfer instructions (page 12) to mark one dragonfly/grass design (page 117) on the center of the panel, beginning 10" (25 cm) above the bottom edge. Mark additional grass blades on either side 5" (13 cm) above the bottom edge. Embroider each cluster of three grass blades in stem stitch using a #8 crewel needle. Work each cluster from left to right, using three strands of grasshopper floss for the first blade, one strand of grasshopper and two strands of parrot green for the second blade, and three strands of parrot green for the third blade.

STEP 2   For the dragonfly, thread a #8 crewel needle with two strands of grape floss and one strand of grape blending filament. Work the dragonfly's body and head in stem stitch. Embroider two 3-wrap French knots for the eyes and two $1/8$" (3 mm) straight stitches for the antennae. Use one strand of lilac floss and two strands of lilac blending filament to outline the wings in stem stitch. Use one strand each of lilac floss, grape blending filament, and lilac blending filament to work the wing details in stem stitch.

Use the template pattern on page 117.

VARIATION

Substitute a bumblebee for the dragonfly and add a perky tulip to the picture for a quick change of scenery! Use the pattern on page 117. You will need the following additional fibers:

■	EBONY DMC #310
■	FOG DMC #762
■	SUNSHINE YELLOW DMC #726
■	GRASSHOPPER DMC #703
■	SILVER METALLIC KREINIK #001

Work the tulip in stem stitch: Use two strands of parrot green floss and one strand of grasshopper floss for the stem. Use three strands of lilac floss for the flower. Outline the bumblebee's head and midsection in stem stitch, using three strands of ebony floss. Work ebony straight stitches, about 1/8" (3 mm) long, for the antennae and legs. Outline the remaining two body sections in stem stitch, using three strands of sunshine yellow floss. Outline the wings in stem stitch, using two strands of fog floss and one strand of silver blending filamanent. Work the wing details in straight stitches.

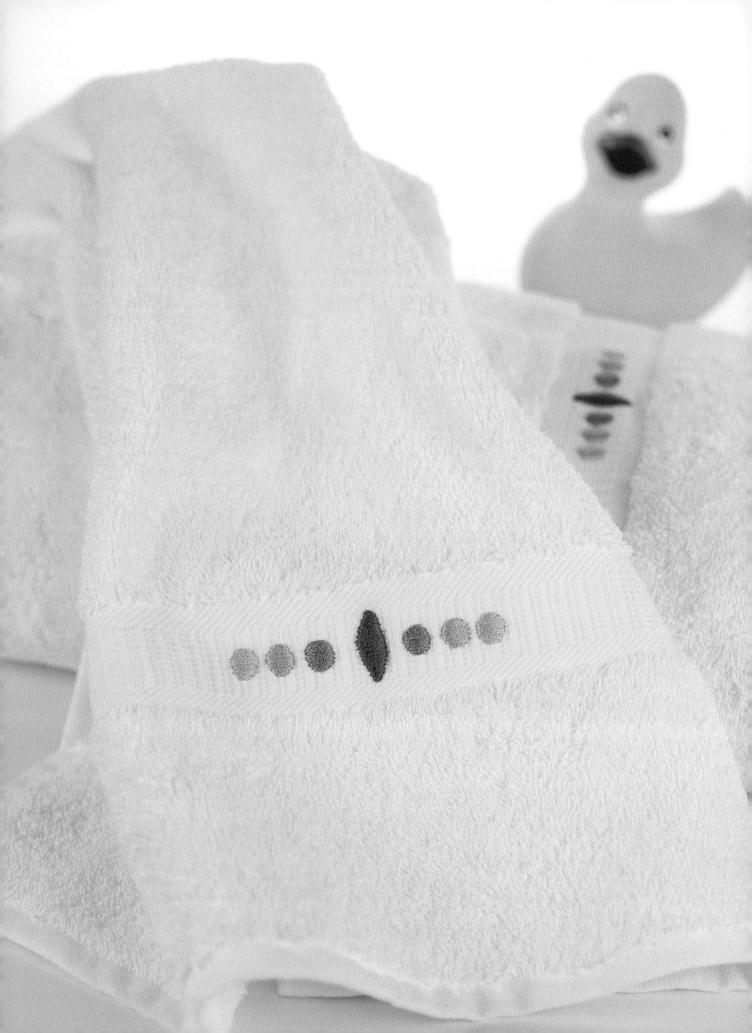

- WHITE COTTON TERRY CLOTH BATH AND HAND TOWELS
- SIX-STRAND COTTON EMBROIDERY FLOSS; WE USED ONE SKEIN EACH OF DMC #958, #959, #964, #720
- #8 CREWEL NEEDLES
- #24 CHENILLE NEEDLES
- EMBROIDERY SCISSORS
- 5" × 7" (13 CM × 18 CM) OVAL EMBROIDERY HOOP
- TAPE MEASURE
- SILK PINS (STRAIGHT PINS WITH GLASS OR METAL HEADS)
- DRESSMAKER'S CARBON PAPER
- STYLUS

STITCHES USED
BACKSTITCH (PAGE 15)
SATIN STITCH (PAGE 18)

 **SKILL LEVEL**
INTERMEDIATE

 **ESTIMATED TIME REQUIRED**
3–4 HOURS FOR BATH TOWEL
1–2 HOURS FOR HAND TOWEL

Bath Towels

Simple geometric shapes make a clean, bold statement on a set of pristine white terry towels. We chose colors reminiscent of the seaside, but the design would be equally arresting in other bright hues. Choose three different shades of floss to achieve the unique "fading out" effect. The possibilities are endless—just use your imagination!

MERMAID GREEN DMC #958

SEAFOAM DMC #959

ICY AQUAMARINE DMC #964

SUNSET ORANGE DMC #720

Bath Towels

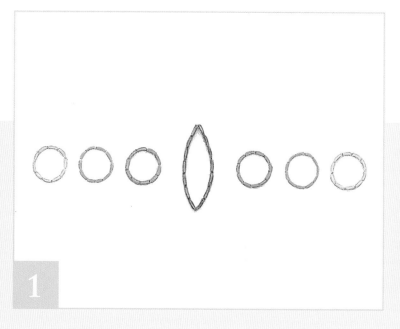

STEP 1 Follow the Dressmaker's Carbon Transfer instructions (page 12) to mark one design motif on the center of each towel's horizontal band. On the bath towel, mark two additional motifs evenly spaced on either side. Thread a #8 crewel needle with two strands of mermaid green floss, and backstitch the two A circles of each motif. Backstitch the B circles with seafoam floss and the C circles with icy aquamarine floss. Backstitch the marquise shape in the middle with sunset orange.

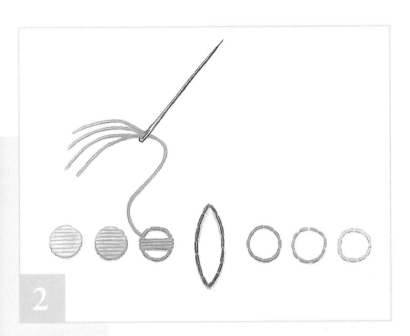

STEP 2 Thread a #24 chenille needle with four strands of icy aquamarine floss. Starting at the widest part of the first C circle, bring the needle and thread to the front of the fabric along the edge of the backstitching, insert the needle into the fabric on the opposite side, and pull through to the back to complete the first stitch. Work satin stitches upward to cover the top half of the circle, and end off. Reemerge at the starting point and satin-stitch the lower half of the circle in the same way. Working from left to right, satin-stitch each shape in the design in the appropriate color floss. Remember to use four strands of floss in a #24 chenille needle for good coverage.

Use the template pattern on page 118.

Use the template pattern on page 118.

VARIATION

Embroider the same design in colors straight from Provence. Three vivid shades of Mediterranean blue team up with sunshine yellow to add a touch of pastoral France to your bath. Run the satin stitching vertically instead of horizontally for a different stitch texture. Here's the palette:

INDIGO DMC #792

PERIWINKLE DMC #793

MEDITERRANEAN SKY DMC #794

SUNSHINE YELLOW DMC #726

- DENIM SHOWER CURTAIN
- SIX-STRAND COTTON EMBROIDERY FLOSS; WE USED THREE SKEINS OF DMC BLANC (ALLOW FOR FOUR), ONE SKEIN OF #725.
- #8 CREWEL NEEDLES
- EMBROIDERY SCISSORS
- 4" (10 CM) ROUND EMBROIDERY HOOP
- SILK PINS (STRAIGHT PINS WITH GLASS OR METAL HEADS)
- DRESSMAKER'S CARBON PAPER
- STYLUS
- IRON AND IRONING BOARD

STITCHES USED
OVERCAST STITCH (PAGE 18)
BUTTONHOLE STITCH (PAGE 20)

 SKILL LEVEL
INTERMEDIATE

 ESTIMATED TIME REQUIRED
1 HOUR PER PATTERN REPEAT

Shower Curtain

☐ WHITE DMC BLANC

▨ SUNFLOWER DMC #725

Cheerful white flowers float across the top of an indigo denim shower curtain

—the perfect prescription for giving your bath a casual, laid-back ambiance.

The easy floral embroidery is done in overcast stitch and buttonhole stitch.

While we used a ready-made shower curtain, the experienced seamstress may

prefer to buy fabric yardage and sew a curtain using a commercial pattern.

Shower Curtain

1

STEP 1 Iron the shower curtain to remove any wrinkles. Follow the Dressmaker's Carbon Transfer instructions (page 12) to mark single flowers in between the ring holes across the top of the curtain, starting about 1 ¹/2" (4 cm) from the edge. Thread a #8 crewel needle with two strands of sunflower floss. Embroider the swirl at the center of each flower in overcast stitch.

VARIATION

The flower design takes on a new look when embroidered in a different color palette. Choose your colors according to the shower curtain background. Against a crisp white cotton shower curtain, peacock blue flowers with vibrant sunshine yellow centers look absolutely stunning. Here's the palette:

PEACOCK BLUE DMC #312

SUNSHINE YELLOW DMC #726

STEP 2 Thread a #8 crewel needle with three strands of white floss. Embroider each flower petal in buttonhole stitch. Tie off after each petal.

Use the template pattern on page 118.

- SILVER VANITY SET
- TWO 9" × 9" (23 CM × 23 CM) SQUARES OF WHITE LINEN FABRIC
- SIX-STRAND COTTON EMBROIDERY FLOSS; WE USED ONE SKEIN OF DMC #800
- #9 CREWEL NEEDLES
- #8 CREWEL NEEDLES
- EMBROIDERY SCISSORS
- 5" × 7" (13 CM × 18 CM) OVAL EMBROIDERY HOOP
- SILK PINS (STRAIGHT PINS WITH GLASS OR METAL HEADS)
- DRESSMAKER'S CARBON PAPER
- STYLUS
- SPRAY STARCH
- IRON AND IRONING BOARD

STITCH USED
DOUBLE BACKSTITCH (PAGE 16)

SKILL LEVEL
ADVANCED

ESTIMATED TIME REQUIRED
4–5 HOURS FOR THE HAND MIRROR
3–4 HOURS FOR THE HAIRBRUSH

 PASTEL BLUE DMC #800

Vanity Set

Any lady would be delighted to add a magnificent vanity set to her dressing table. This design marries lavish scrolls and swirls with classically simple circle and diamond motifs. All are embroidered in delicate pastel blue on a white linen ground. The finished embroideries are mounted in a silver brush and mirror set for a truly timeless heirloom. Double backstitch is the only stitch used in this project, making it a concentrated study in shadow work embroidery.

vanity Set

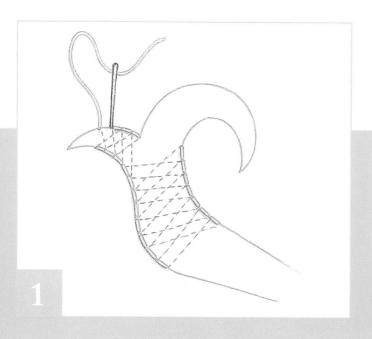

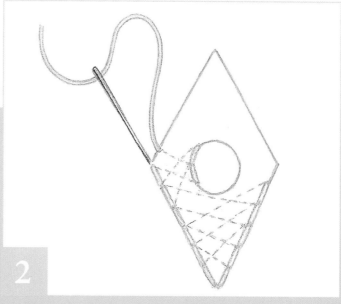

STEP 1 Apply a light coat of spray starch to one square of linen fabric. Allow two or three minutes for the fabric to absorb the starch, then iron the fabric dry. Follow the Direct Tracing Transfer instructions (page 12) to mark the mirror or brush scroll design (page 118) on the fabric. Thread a #9 crewel needle with two strands of pastel blue floss. Begin the shadow work (double backstitch) embroidery on the scroll part of the design, starting at any outside point. Work up the left arm of the V until you reach the place where the design forks into two arches. Stitch the left arch first, remembering to make the stitches on the inside curve slightly smaller/shorter than those on the outside curve. Work the right arch the same way. Weave in the ends on the underside to "shadow" the small open triangle that occurs at the fork. Continue until all the scrollwork is embroidered.

STEP 2 For the diamonds, thread a #9 crewel needle with two strands of pastel blue floss. Begin the double backstitch at the bottom point of the diamond. When you reach the circle, work up toward the left, around the circle, ending at the top of the circle. Weave the needle through the crisscrossed threads on the underside, and reemerge at right side edge where the stitching left off. Work up the right side in the same way and then work both sides to the top point.

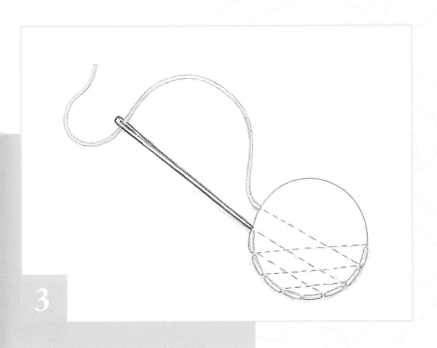

STEP 3   For the circles, thread a #8 crewel needle with two strands of pastel blue floss. Make one or more backstitches at the bottom edge of the circle. When the stitches begin to arc upward, begin the side-to-side motion of the double backstitch. Finish off the circle as you started, by making a few backstitches.

Use the template patterns on page 118.

VARIATION

A classic vanity set that can be passed down to future generations is truly a keepsake treasure. Other "icy" pastels, such as pale pink or lavender, would also look elegant against the silverplating. White-on-white is another palette option that never falls out of style, and ecru-on-white looks stunning too!

Sachet Pouch

Perfect for your best friend's birthday, for Mom on Mother's Day, or as a bride's keepsake gift to her wedding attendants, this simple pouch is easy to whip up in advance of special occasions.

Begin with two 5 " × 10 " (13 cm × 25 cm) strips of pink linen fabric. Fold each strip in half, right side out, so it is square, and finger-press the fold. Unfold one strip and mark the design (page 119) on one square only, using the Direct Tracing Transfer method (page 12). Mount the marked strip in a scroll frame. Embroider each corner scroll design in stem stitch, using two strands of sand floss in a #9 crewel needle.

To construct the pouch, refold both strips, right side out, and press to set the creases. Pin the squares together, embroidery on the inside and raw edges matching. Machine-stitch 1/2 " (1 cm) from the raw edges around three sides. Trim the seam allowance to 1/4 " (5 mm), clip the corners, and zigzag the raw edges. Turn right side out. Hand-sew ribbon ties to the folded edges in pairs. Tuck potpourri inside the pouch, and tie the ribbons closed.

STITCH USED
STEM STITCH (PAGE 15)

SAND DMC #644

Picture Frame

STITCH USED
BACKSTITCH (PAGE 15)

ULTRAMARINE BLUE DMC #825

SNOW WHITE DMC #5200

Whimsical spirals create an eye-catching border around a favorite photograph. Choose thread colors that coordinate with your photo, as we did for this outdoor scene. With motifs this simple, even beginners can create spectacular results.

Start with a 12" × 12" (30 cm × 30 cm) square of raw linen fabric. Transfer the design (page 119) using the Dressmaker's Carbon Transfer method (page 12). Mount the fabric in an oval embroidery hoop. Backstitch the box-shaped spirals first, using two strands of ultramarine blue floss in a #9 crewel needle. Continue all the way around. Then backstitch the circular spirals using two strands of snow white floss.

A professional framer finished our picture frame, but if you have some framing experience, you may wish to complete this step yourself. Purchase an appropriately sized frame and mat for your standard-sized photo. Mount the embroidered fabric on the mat with spray adhesive, making sure it is perfectly centered. Use a sharp mat knife on a protected surface to trim away the excess fabric from the mat opening and around the edges. Mount your photo and the mat in the frame.

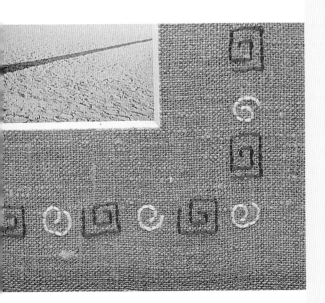

Hall Mirror

STITCH USED
STEM STITCH (PAGE 15)

FOREST GLADE DMC #3052

WOODLAND MEADOW DMC #3051

Transform an ordinary mirror into a needlework showcase. This handsome project is worked entirely in stem stitch. Subtle shades of green and an elegant satin ribbon hanger complement the walnut frame.

Stiffen a 5" × 7" (13 cm × 18 cm) piece of white linen fabric with iron-on spray starch. Transfer the scroll design (page 119) to the linen using the Direct Tracing Transfer method (page 12). Mount the fabric in an oval hoop so that the design is centered inside the frame. Embroider the four upper scroll motifs in stem stitch, using two strands of forest glade floss in a #9 crewel needle. Repeat to embroider the lower half. Use one strand each of forest glade floss and woodland meadow floss to work the motifs in the middle. Mount the finished embroidery in the display area of the mirror frame.

Crystal Vanity Jar

STITCHES USED
STEM STITCH (PAGE 15)
DETACHED CHAIN STITCH
(A.K.A. LAZY DAISY) (PAGE 20)
FRENCH KNOT (PAGE 20)
BULLION STITCH (PAGE 21)

PASTEL GREEN DMC #955

PASTEL PINK DMC #3713

BABY BLUE DMC #3325

ROSEBUD DMC #761

LEMON DROP DMC #3078

The beauty of the garden is captured in this circle of delicate flowers designed especially for the lid of a hand-cut lead crystal jar.

Stiffen a 6" × 6" (15 cm × 15 cm) square of white linen fabric with iron-on spray starch. Transfer the floral design (page 120) to the linen using the Direct Tracing Transfer method (page 12). Mount the fabric in a round hoop so that the design is centered inside the frame. Embroider the foliage first, using two strands of pastel green floss in a #9 crewel needle. Work the stems in stem stitch and the leaves in detached chain stitch. Change to two strands of pastel pink. Embroider the daisies in detached chain stitch, tying off after each group of three, or after each single daisy, as appropriate. Thread a #7 betweens needle with two strands of baby blue floss. For each forget-me-not, embroider four 2-wrap French knots as the petals. Tie off after each flower. Thread a #8 milliner's needle with two strands of rosebud floss. Work two 8-wrap bullion stitches side by side to create each bullion rosebud. Tie off after each cluster. For each rose, first work two 5-wrap bullion stitches side by side for the rose center. Change to one strand each of pastel pink and rosebud floss. Embroider three 9-wrap bullion stitches around each rose center, followed by six 10-wrap bullion stitches. Tie off after completing each rose. Thread a #7 betweens needle with two strands of lemon drop floss. Embroider a 2-wrap French knot at the center of each lazy daisy and forget-me-not. Additional leaves may be added in detached chain stitch if desired.

Trinket Box

Work these cheerful flowers in the tantalizing colors of a tropical sunset—you'll feel like you're in an embroiderer's paradise.

Stiffen a 7" × 9" (18 cm × 23 cm) piece of white linen fabric using iron-on spray starch. Transfer the design (page 120) to the linen using the Direct Tracing Transfer method (page 12). Using two strands of orange blaze floss in a #9 crewel needle, embroider the left daisy's petals in chain stitch. In the same way, embroider the middle daisy with lemon sunshine floss, and the right daisy with turquoise floss. Chain-stitch the center of the middle daisy in orange blaze floss and the centers of the other two daisies in lemon sunshine floss. Tie off after each color change. Work the zigzag borders in interlaced backstitch. Backstitch the outer zigzag lines with two strands of turquoise floss and the inner zigzags with two strands of orange blaze floss. Use a #26 tapestry needle to weave two strands of lemon sunshine floss in and out of the backstitching.

STITCHES USED
CHAIN STITCH (PAGE 19)
INTERLACED BACKSTITCH (PAGE 16)

■ TURQUOISE DMC #806

■ ORANGE BLAZE DMC #947

■ LEMON SUNSHINE DMC #973

Trivet

STITCHES USED
STEM STITCH (PAGE 15)
STRAIGHT STITCH (PAGE 16)

VALENCIA DMC #741

CITRON DMC #742

DARK OLIVE DMC #3011

OLIVE DMC #3012

A lovely citrus design, embroidered on a ready-made linen trivet, protects your dining table with style. This project is perfect for the embroiderer who needs a quick gift or for any stitcher who craves instant gratification. In fact, you can whip up an entire set in a weekend!

Remove the padding from inside the trivet. Press the linen cover with a warm iron to remove any wrinkles. Transfer the citrus design (page 121) to the linen cover using the Dressmaker's Carbon Transfer method (page 12). Mount the fabric in a 3" (8 cm) round hoop so that the design is centered inside the frame. (Make sure that the inner ring is inside the trivet pocket and that only the top layer of fabric is framed up.) Thread a #9 crewel needle with one strand each of Valencia and citron floss. Working from left to right, outline each orange in stem stitch, and tie off. Change to two strands of Valencia floss. Work the curved lines on each orange in stem stitch and the tiny angular details in straight stitch. Change to one strand each of dark olive and olive floss. Beginning at the lower left, embroider the outline and center vein of each leaf in stem stitch. Tie off after each leaf (or leaf pair) rather than carrying the thread across the back. Reinsert the padding before using the trivet.

Guest Towel

Stitch an enchanting floral spray in splashy shades of azalea, lemon drop, cherry red, orange blaze, and key lime for a stunning twist on an old classic. That's right, this timeless embroidery design takes on a whole new look when worked in snazzy brights.

Begin by stiffening a ready-made white linen guest towel with iron-on spray starch. Use the Direct Tracing Transfer method (page 12) to transfer the design (page 121), centering it on the towel about 5" (13 cm) above the lower edge. Thread a #9 crewel needle with two strands of key lime floss. Embroider the five stems in stem stitch and the leaves in detached chain stitch. Change to two strands of orange blaze floss. Work the daisy petals in detached chain stitch, tying off after each flower. Thread a #7 betweens needle with two strands of cherry red floss. Embroider each posey petal with a 2-wrap French knot, tying off after each flower. Thread a #8 milliners needle with two strands of azalea floss. Work two 8-wrap bullion stitches side by side four times to create the rosebuds in the center of the design. Thread a #9 crewel needle with two strands of key lime floss, and make two lazy daisy leaves at the base of each rosebud. Thread a #7 betweens needle with two strands of lemon drop floss, and work a 2-wrap French knot at the center of each daisy and small red posey. Machine-stitch a length of cherry red rickrack to the towel 1" (3 cm) above the lower edge.

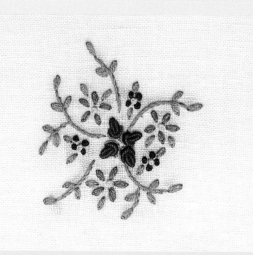

STITCHES USED
STEM STITCH (PAGE 15)
DETACHED CHAIN STITCH
    (A.K.A. LAZY DAISY) (PAGE 19)
FRENCH KNOT (PAGE 20)
BULLION STITCH (PAGE 21)

KEY LIME DMC #906

ORANGE BLAZE DMC #947

CHERRY RED DMC #321

AZALEA DMC #3804

LEMON DROP DMC #3078

cocktail Napkins

STITCH USED
OVERCAST STITCH (PAGE 18)

 SAND DMC #644

Chic cocktail napkins are classically simple, and they're a breeze to embroider, too! One stitch, one thread color, purchased hemstitched linen squares—what could be easier?

Stiffen a white linen cocktail napkin with iron-on spray starch. Transfer the scroll design (page 121) to one corner of the napkin fabric using the Direct Tracing Transfer method (page 12). To mount the napkin in a hoop so the design is centered, follow the tip for making a false muslin edge (page 14). Thread a #9 crewel needle with two strands of sand floss. Working from left to right, embroider the scrolls in overcast stitch, and tie off. Remove the muslin from the corner of the napkin and discard it. Repeat to embroider each napkin in the set.

Safeguard your precious photos and memorabilia in a monogrammed

album—there's just no better way to commemorate life's milestones. Better

yet, start a tradition by making an album for each of life's celebrations—

births, graduations, weddings, anniversaries.

Choose a commercial pattern and raw linen fabric for your album cover. Cut

out the fabric as the pattern directs, and then transfer the heart design (page

122) to the front cover area using the Dressmaker's Carbon Transfer method

(page 12). Thread #9 crewel needle with two strands of lilac floss. Mount

the fabric on a scroll frame. Beginning at the bottom, work up the left side

of the heart shape in chain stitch. Then work up the right side. To add a

monogram, use the alphabet on page 125. Draft a 2 1/2" (6 cm) line on

tracing paper, center the line horizontally on the desired letter, and trace.

Use dressmaker's carbon to transfer the monogram to the linen within the

embroidered heart. Embroider the monogram in chain stitch, too. Follow

the pattern instructions to complete the album cover.

Memory Album

STITCH USED
CHAIN STITCH (PAGE 19)

LILAC DMC #209

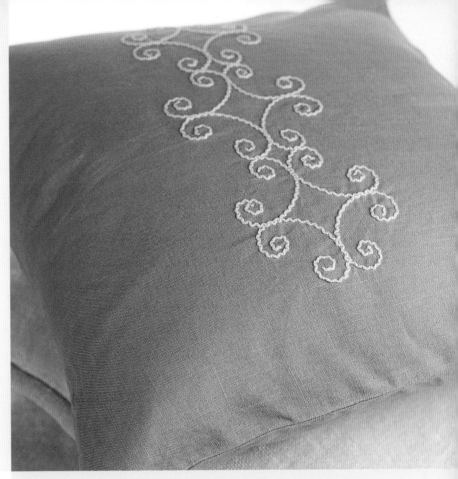

Square Sofa Pillow

STITCH USED
INTERLACED BACKSTITCH (PAGE 16)

WHITE SILK & IVORY YARN #02

To create a one-of-a-kind accessory for your sofa, embellish a purchased linen pillow cover with decorative stitching. This entire design is worked in interlaced backstitch, a combination stitch that is usually worked in two colors. Here, a soft white yarn is used for both parts of the stitch, placing the accent on texture instead.

Unbutton the cover, and remove the pillow insert. Using the Dressmaker's Carbon Transfer method (page 12), mark the scroll design (page 122) three times down the middle of the pillow cover (mark the center design first). Thread a #24 chenille needle with one strand of white yarn. Mount the pillow cover in a 6" (15 cm) round embroidery hoop so the top motif is showing. Starting in the lower left quadrant and working clockwise, embroider each scroll line with 1/8" (3 mm) backstitches, and tie off. Work the two remaining design repeats in the same manner. Use a #28 tapestry needle to weave one strand of white yarn in and out of the backstitches. Press the cover with a warm iron before reinserting the pillow form.

Ladybug Place Mat

This little ladybug, embroidered on a place mat, looks like she's just wandered in from the garden. For an amusing twist, change her location on each place mat you embroider.

Mark the design (page 122) using the Dressmaker's Carbon Transfer method (page 12). Thread a #24 chenille needle with three strands of cherry red floss. Work the ladybug's entire body in satin stitch, and tie off. Change to three strands of ebony floss. Satin-stitch the seven spots on the ladybug's body (see the template diagram), and tie off. Work her head in satin stitch, too. Thread a #8 crewel needle with two strands of ebony floss. Using tiny straight stitches, work the antennae and legs.

STITCHES USED
STRAIGHT STITCH (PAGE 16)
SATIN STITCH (PAGE 18)

 CHERRY RED DMC #321

 EBONY DMC #310

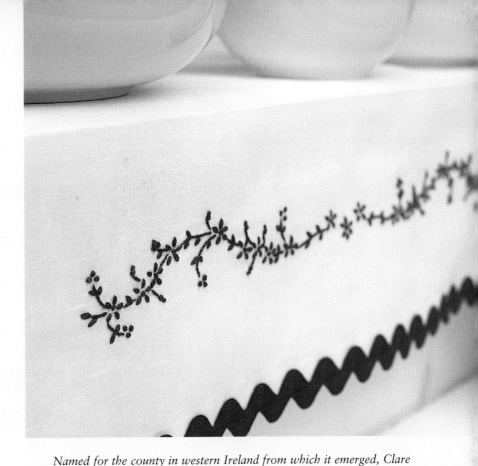

Shelf Cloth

STITCHES USED
STEM STITCH (PAGE 15)
DETACHED CHAIN STITCH
    (A.K.A. LAZY DAISY) (PAGE 19)
FRENCH KNOT (PAGE 20)

 COBALT BLUE DMC #797

Named for the county in western Ireland from which it emerged, Clare embroidery is a unique style that dates back to the late nineteenth century. It is traditionally worked on sturdy cotton poplin fabric and has been used to adorn everything from children's clothing to household linens. Make our crisp shelf cloth to add a charming touch o' the Emerald Isle to your home!

You'll need a strip of white cotton fabric, 10" (25 cm) wide and cut 5" (13 cm) shorter than the shelf length. Machine-sew a shirttail hem on all four edges. Stiffen the fabric using iron-on spray starch. Make two mirror-image copies of the floral template pattern (page 123). Tape both halves together, making sure the six-petaled flower is in the middle. Following the Direct Tracing Transfer instructions (page 12), mark the design on the middle of the strip, 3" (8 cm) above the long lower edge. Work the entire design using two strands of cobalt blue floss in a #9 crewel needle. Work the vines and stems in stem stitch and the lazy daisy flowers and leaves in detached chain stitch. Use 2-wrap French knots for the flower centers and 3-wrap French knots for the "dot" flowers. Machine-sew cobalt blue rickrack along the lower edge.

Tray Cloth

Transform an ordinary place mat into an elegant tray cloth. Our exquisite hydrangea blossoms are sprinkled inside a scalloped border on a ground of ivory-colored linen. Simple yet stunning!

Use the Direct Tracing Transfer method (page 12) to mark the scalloped border design (page 123) on the place mat. Work out the full size on tracing paper first so the scallop falls 1 1/8" (3 cm) in from the edge all around. Mark the hydrangea blossoms inside the border in a loose grid arrangement. Thread a #9 crewel needle with two strands of cornflower floss. Secure a corner of the place mat in an oval embroidery hoop. Work the scalloped border in chain stitch all around, moving the hoop as necessary. For the hydrangeas, change to a 3" (8 cm) round hoop. Using one strand each of leaf green and spring green floss, work the hydrangea stem in stem stitch and the leaves in detached chain stitch. For the blossoms, thread a #7 betweens needle with one strand each of cornflower and arctic blue floss. Embroider a 2-wrap French knot for each blossom to form the flower head.

STITCHES USED

STEM STITCH (PAGE 15)
DETACHED CHAIN STITCH
  (A.K.A. LAZY DAISY) (PAGE 19)
CHAIN STITCH (PAGE 19)
FRENCH KNOT (PAGE 20)

CORNFLOWER DMC #341

SEA GREEN DMC #772

SPRING GREEN DMC #3348

ARCTIC BLUE DMC #3753

Nightgown Case

The old European custom of keeping your nightclothes on top of the bed in a decoratively embroidered case has all but disappeared—until now. Revive an old-fashioned tradition when you make our redwork nightgown case. A contemporary butterfly silhouette updates the classic design.

You'll need 2 yards (1.8 meter) of white cotton fabric, 1 yard (.9 meter) of jumbo rickrack, and sewing notions. Cut off a 1/2-yard piece (1/2-meter) from the fabric, and set the rest aside. Mark the butterfly flap design (page 124) on the 1/2-yard (1/2-meter) piece using the Dressmaker's Carbon Transfer method (page 12). Thread a #9 crewel needle with two strands of cherry red floss. Using an oval hoop, embroider the dots around the edge in padded satin stitch. Tie off after each dot. For the butterfly, use a 6" (15 cm) round hoop. Work the wings and body in padded satin stitch and the antennae in stem stitch. Complete each antenna with a satin stitch dot.

To make the case, cut one front piece, 15 1/2" × 13 7/8" (39 cm × 36 cm), and one back piece, 13 7/8" × 13 7/8" (36 cm × 36 cm), from the reserved fabric. Cut out the embroidered flap on the marked line. Also cut a matching flap lining. Fold under one long edge of the case front, sewing a 2" hem. Sew the front and back together, right sides facing, around three sides. Clip the corners diagonally, turn right side out, and press with a warm iron. Press the straight edge of the flap lining 5/8" (1 cm) to the wrong side. Baste red rickrack to the right side of the embroidered flap along the seam line, clipping at the point if necessary. Sew the flaps right sides together along the basting line, clip as needed, and turn right side out. Sew the flap to the case back, right sides together and raw edges matching. Press the seam allowances to the inside, and slip-stitch the lining closed along the folded edge. Finish by edge-stitching the flap.

STITCHES USED
STEM STITCH (PAGE 15)
PADDED SATIN STITCH (PAGE 18)

 CHERRY RED DMC #321

Template Patterns

To work any of the embroidery projects in this book, you will need to use a template pattern. A template is a guide for the stitching that is transferred directly to the embroidery fabric. A description of three different template pattern transfer methods can be found on pages 12-13.

To enlarge a pattern for transferring, use a photocopy machine and set it to the percentage as indicated on the pattern page.

Before transferring the design to fabric, read through the project instructions and study the illustrations to be sure you understand the procedures required to successfully complete the piece. Be sure to match the alignment markings on the patterns (represented by dashed lines) with the straight grain of the fabric before you begin marking. If you have questions or run into difficulty, ask your local needlework shop owner or one of your stitching friends for some help. Refer to the Resource Guide (page 126) for additional assistance.

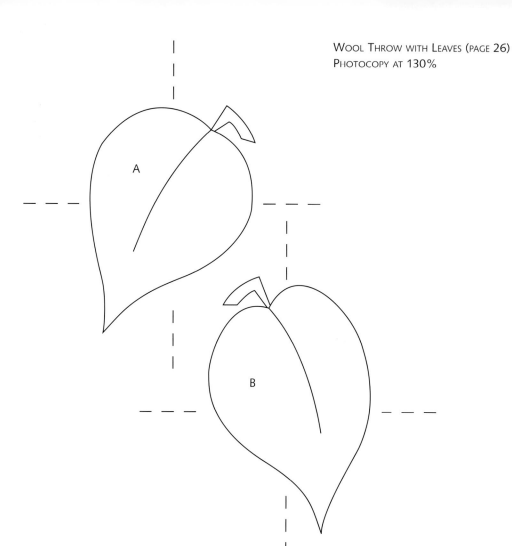

WOOL THROW WITH LEAVES (PAGE 26)
PHOTOCOPY AT 130%

OVERLAP MIRROR-IMAGE PHOTOCOPY AND TAPE HERE

OVERLAP MIRROR-IMAGE PHOTOCOPY AND TAPE HERE

SILK SOFA PILLOW (PAGE 30)
PHOTOCOPY AT 155%

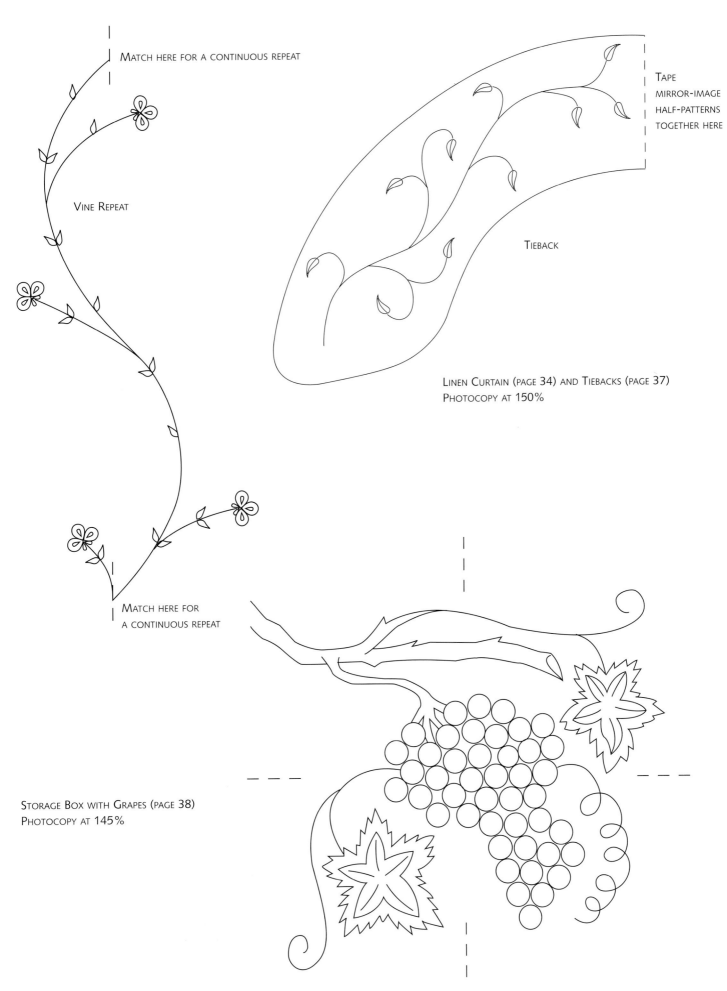

MATCH HERE FOR A CONTINUOUS REPEAT

TAPE
MIRROR-IMAGE
HALF-PATTERNS
TOGETHER HERE

VINE REPEAT

TIEBACK

LINEN CURTAIN (PAGE 34) AND TIEBACKS (PAGE 37)
PHOTOCOPY AT 150%

MATCH HERE FOR
A CONTINUOUS REPEAT

STORAGE BOX WITH GRAPES (PAGE 38)
PHOTOCOPY AT 145%

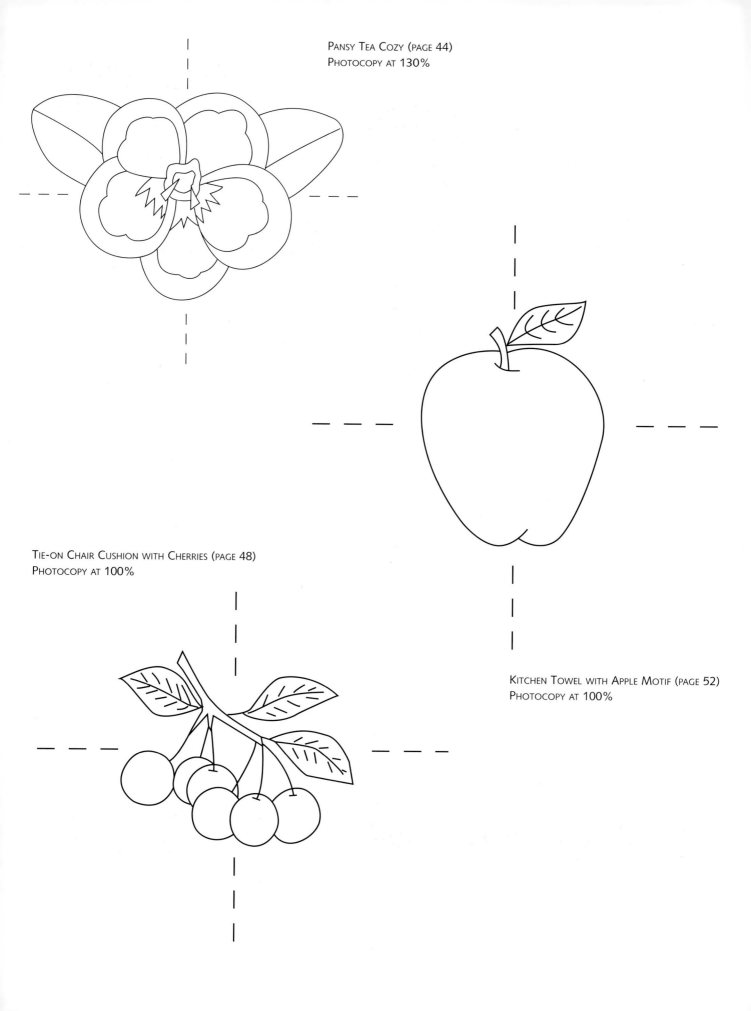

PANSY TEA COZY (PAGE 44)
PHOTOCOPY AT 130%

TIE-ON CHAIR CUSHION WITH CHERRIES (PAGE 48)
PHOTOCOPY AT 100%

KITCHEN TOWEL WITH APPLE MOTIF (PAGE 52)
PHOTOCOPY AT 100%

SMALL SCROLL (NAPKIN)
PHOTOCOPY AT 100%

SCROLLWORK TABLECLOTH AND NAPKINS (PAGE 56)

LARGE SCROLL (TABLECLOTH)
PHOTOCOPY AT 100%

BED LINENS (PAGE 62)
PHOTOCOPY AT 100%

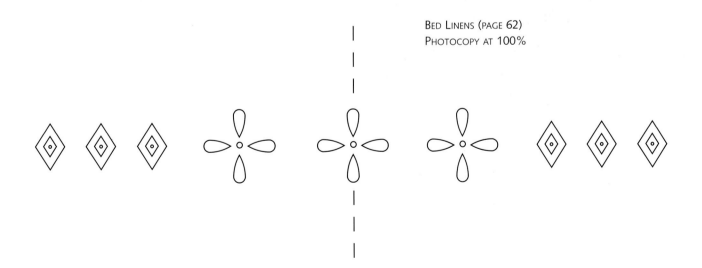

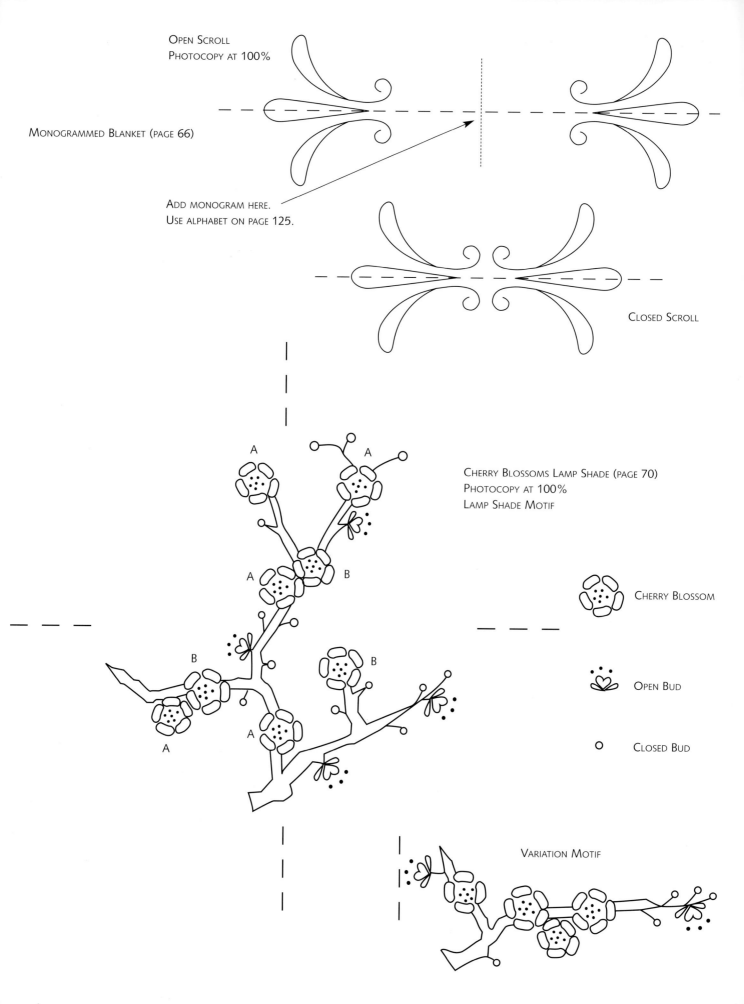

OPEN SCROLL
PHOTOCOPY AT 100%

MONOGRAMMED BLANKET (PAGE 66)

ADD MONOGRAM HERE.
USE ALPHABET ON PAGE 125.

CLOSED SCROLL

CHERRY BLOSSOMS LAMP SHADE (PAGE 70)
PHOTOCOPY AT 100%
LAMP SHADE MOTIF

A

A

A

B

B

B

A

CHERRY BLOSSOM

OPEN BUD

CLOSED BUD

VARIATION MOTIF

FOLD HERE

SACHET POUCH (PAGE 96)
PHOTOCOPY AT 100%

PICTURE FRAME (PAGE 97)
PHOTOCOPY AT 100%

HALL MIRROR (PAGE 100)
PHOTOCOPY AT 130%

CRYSTAL VANITY JAR (PAGE 99)
PHOTOCOPY AT 110%

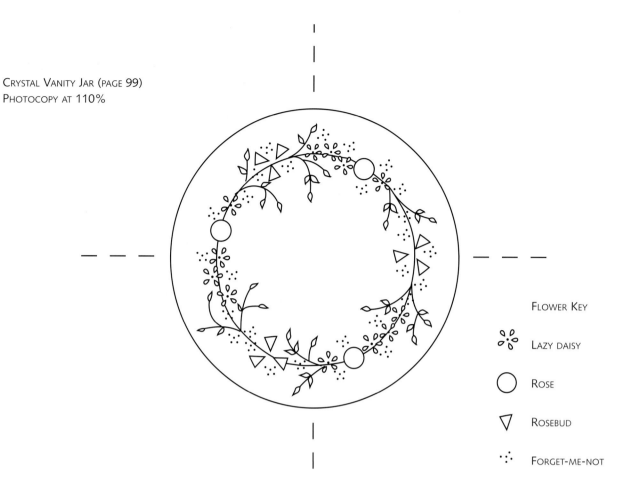

FLOWER KEY

LAZY DAISY

ROSE

ROSEBUD

FORGET-ME-NOT

TRINKET BOX (PAGE 100)
PHOTOCOPY AT 100%

TRIVET (PAGE 101)
PHOTOCOPY AT 100%

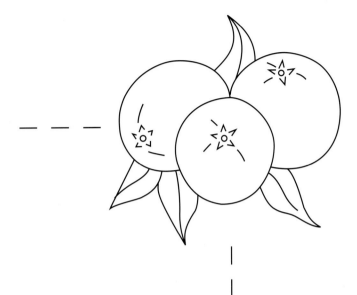

GUEST TOWEL (PAGE 102)
PHOTOCOPY AT 100%

FLOWER KEY

 ROSEBUD

 POSEY

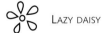 LAZY DAISY

COCKTAIL NAPKINS (PAGE 103)
PHOTOCOPY AT 100%

SQUARE SOFA PILLOW (PAGE 105)
PHOTOCOPY AT 140%

X
ADD MONOGRAM HERE.
USE ALPHABET ON PAGE 125.

LADYBUG PLACE MAT (PAGE 106)
PHOTOCOPY AT 100%

MEMORY ALBUM (PAGE 104)
PHOTOCOPY AT 130%

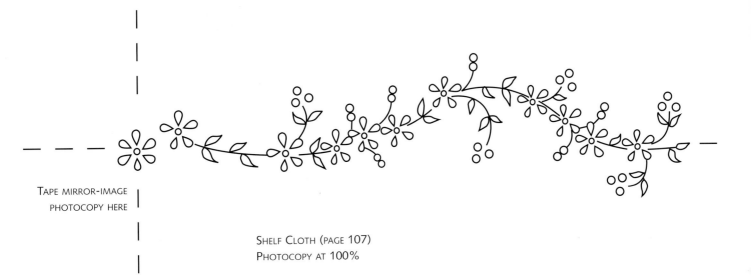

TAPE MIRROR-IMAGE
PHOTOCOPY HERE

SHELF CLOTH (PAGE 107)
PHOTOCOPY AT 100%

TRAY CLOTH (PAGE 108)
PHOTOCOPY AT 130%

NIGHTGOWN CASE (PAGE 109)
PHOTOCOPY AT 145%

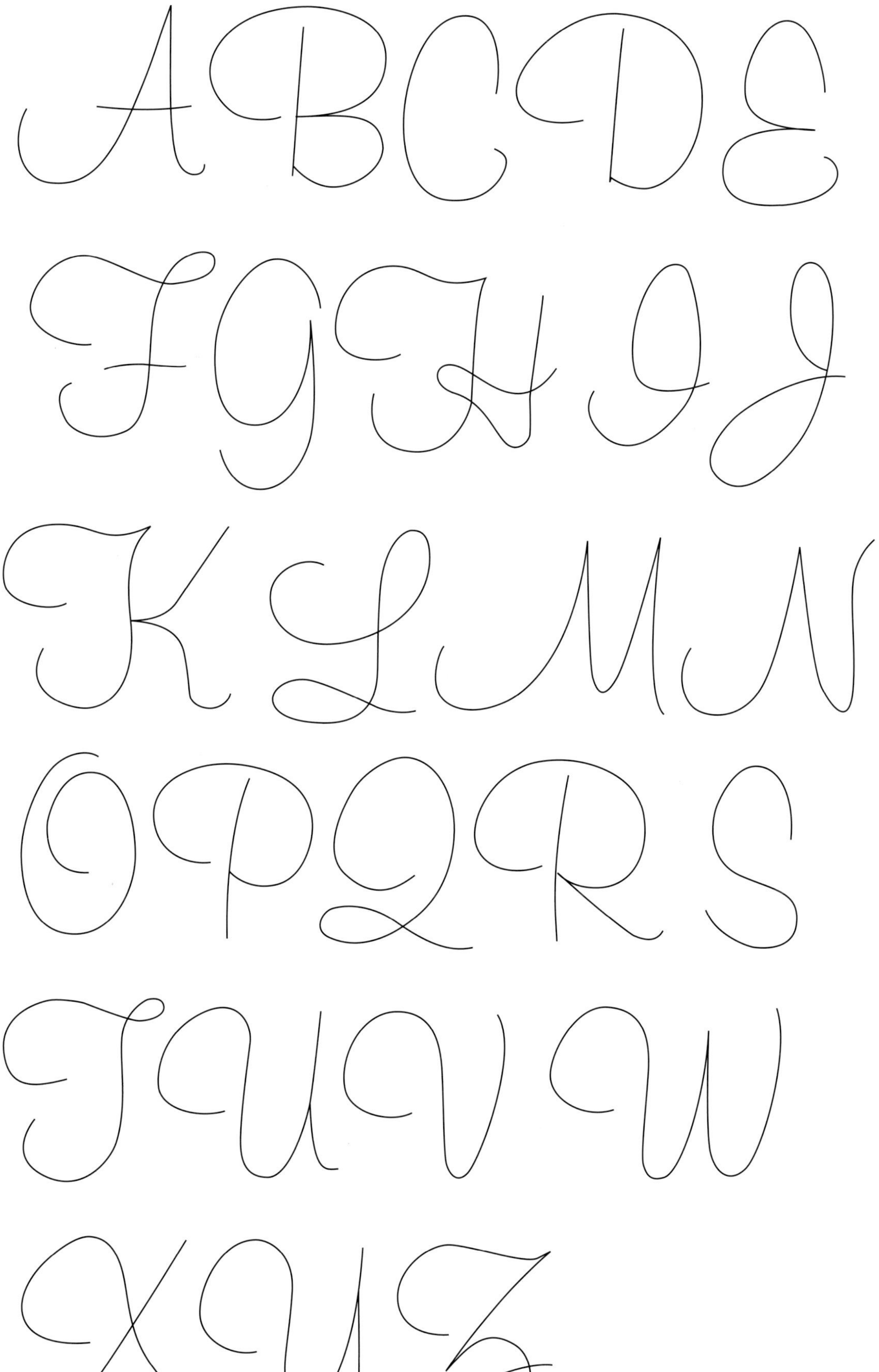

A B C D E

F G H I J

K L M N

O P Q R S

T U V W

X Y Z

MONOGRAM ALPHABET
PHOTOCOPY AT 125%

Resource Guide

Ellen Moore Johnson Designs
3806 Somerset Place
Tuscaloosa, AL 35405-2757
(888) 248-4743
http://www.heirloomembroidery.com
Embroidery patterns, embroidery needles,
instructional video tapes, fine fabrics, and
ready-made linens for embroidery projects
(tea cozy, kitchen towels, tablecloth and
napkins, bed linens, trivet, guest towel
and cocktail napkin; fabrics for storage box,
vanity set, sachet pouch, picture frame,
hall mirror, crystal jar, trinket box,
memory album cover, curtain tieback)

Brown Paper Packages
18 Grand Lake
Ft. Thomas, KY 41075
Fine fibers for needlework
(Merino/silk yarn)

Creative Furnishings
12357 Saraglen Drive
Saratoga, CA 95070
(408) 996-7745
Finely handcrafted wood products designed
especially for display of needlework
(mahogany storage box)

The DMC Corporation
South Hackensack Avenue
Port Kearny Building 10A
South Kearny, NJ
07032-4688
(973) 589-0606
http://www.dmc-usa.com

Fine embroidery fibers

(six-strand cotton embroidery floss
and broder medici)

Garnet Hill
231 Main Street
Franconia, NH 03580
(800) 622-6216
http://www.garnethill.com
High-quality bed linens, pillows, curtains,
furniture, and clothing made of natural
fibers
(Merino wool blanket)

Gay Bowles Sales, Inc.
P. O. Box 1060
Janesville, WI 53547
(800) 356-9438
http://www.millhill.com
Distributor of Framecraft products
(lead crystal jar and vanity set)

Hancock Fabrics
2605A West Main Street
Tupelo, MS 38801
(662) 844-7368
http://www.hancockfabrics.com
Patterns, fabrics, sewing notions
(silk dupioni lamp shade fabric, lamp shade
trim, cotton fabric for nightgown case and
shelf cloth)

Kreinik Manufacturing Co., Inc.
3106 Timanus Lane
Suite 101
Baltimore, MD 21244
(410) 281-0040
http://www.kreinik.com
Fine embroidery fibers
(metallic blending filament)

L. L. Bean
Freeport, ME 04033
(800) 221-4221
http://www.llbean.com
Catalog with a variety of linens
for the home
(wool throw)

Pottery Barn
Mail Order Department
P. O. Box 7044
San Francisco, CA 94120-7044
(800) 922-9934 (call for the store
nearest you.)
http://www.potterybarn.com
Bed, bath, and kitchen linens, also window
treatments and throws
(silk sofa cushion, linen curtain, cotton bed
skirt, cotton bath towels, square sofa pillow,
ladybug place mat)

Sudberry House
12 Colton Road
East Lyme, CT 06333
(800) 243-2607
http://www.sudberry.com
Fine wood accessories for needlework
(hall mirror, trinket box)

The Embroiderers' Guild of America
335 West Broadway
Suite 100
Louisville, KY 40202
(502) 589-6956
http://www.egausa.org
Nonprofit organization dedicated to the
advancement of embroidery

Embroidery Floss Color Conversions

NUMERICAL

DMC	COLOR NAME	ANCHOR
#B5200	snow white	#1
Blanc	white	#2
Ecru	ecru	#387
#208	grape	#110
#209	lilac	#109
#310	ebony	#403
#312	peacock blue	#979
#320	dark leaf green	#215
#321	cherry red	#9046
#341	cornflower	#117
#368	leaf green	#214
#420	chestnut	#374
#434	light brown	#310
#498	dark red	#1005
#550	purple	#102
#552	violet	#99
#554	lavender	#96
#597	aquamarine	#1064
#598	icy aqua	#1062
#644	sand	#830
#702	parrot green	#226
#703	grasshopper	#238
#720	sunset orange	#326
#725	canary yellow	#305
#726	sunshine yellow	#295
#727	buttercup	#293
#741	Valencia	#304
#742	citron	#303
#761	rosebud	#1021
#762	fog	#234
#772	sea green	#259
#792	indigo	#941
#793	periwinkle	#176
#794	Mediterranean sky	#175
#797	cobalt blue	#132
#800	pastel blue	#144
#806	turquoise	#169
#815	crimson	#43
#818	cotton candy	#23
#825	ultramarine blue	#162
#841	taupe	#1082
#869	brown	#944
#905	grass	#257
#906	key lime	#256
#947	orange blaze	#330
#955	pastel green	#206
#958	mermaid green	#187
#959	seafoam	#186
#964	icy aquamarine	#185
#972	sunflower	#298
#973	lemon sunshine	#297
#988	green apple	#243
#3011	dark olive	#846
#3012	olive	#844
#3046	wheat	#887
#3051	woodland meadow	#681
#3052	forest glade	#262
#3078	lemon drop	#292
#3325	baby blue	#129
#3326	rose blush	#36
#3346	lime peel green	#267
#3348	spring green	#264
#3362	jade	#263
#3364	moss	#260
#3713	pastel pink	#1020
#3753	arctic blue	#1031
#3804	azalea	#63
#3863	cocoa	#379

ALPHABETICAL

COLOR NAME	DMC	ANCHOR
aquamarine	#597	#1064
arctic blue	#3753	#1031
azalea	#3804	#63
baby blue	#3325	#129
brown	#869	#944
buttercup	#727	#293
canary yellow	#725	#305
cherry red	#321	#9046
chestnut	#420	#374
citron	#742	#303
cobalt blue	#797	#132
cocoa	#3863	#379
cornflower	#341	#117
cotton candy	#818	#23
crimson	#815	#43
dark leaf green	#320	#215
dark olive	#3011	#846
dark red	#498	#1005
ebony	#310	#403
ecru	ecru	#387
fog	#762	#234
forest glade	#3052	#262
grape	#208	#110
grass	#905	#257
grasshopper	#703	#238
green apple	#988	#243
icy aqua	#598	#1062
icy aquamarine	#964	#185
indigo	#792	#941
jade	#3362	#263
key lime	#906	#256
lavender	#554	#96
leaf green	#368	#214
lemon drop	#3078	#292
lemon sunshine	#973	#297
light brown	#434	#310
lilac	#209	#109
lime peel green	#3346	#267
Mediterranean sky	#794	#175
mermaid green	#958	#187
moss	#3364	#260
olive	#3012	#844
orange blaze	#947	#330
parrot green	#702	#226
pastel blue	#800	#144
pastel green	#955	#206
pastel pink	#3713	#1020
peacock blue	#312	#979
periwinkle	#793	#176
purple	#550	#102
rose blush	#3326	#36
rosebud	#761	#1021
sand	#644	#830
sea green	#772	#259
seafoam	#959	#186
snow white	#B5200	#1
spring green	#3348	#264
sunflower	#972	#298
sunset orange	#720	#326
sunshine yellow	#726	#295
taupe	#841	#1082
turquoise	#806	#169
ultramarine blue	#825	#162
Valencia	#741	#304
violet	#552	#99
wheat	#3046	#887
white	Blanc	#2
woodland meadow	#3051	#681

Acknowledgments

You would not be holding this book in your hands if the wonderful people listed below had not been so gracious in sharing their incredible gifts and talents. I will be forever grateful:

- to Mary Ann Hall for her support of my work from the very beginning. Without her help, I would not have had the opportunity to write this book. Thank you, Mary Ann, for encouraging me to share my love of embroidery with others.

- to Martha Wetherill and Shawna Mullen. Thank you for inviting me to write *The Embroidered Home*.

- to Linda Clark for so graciously allowing me to haunt her beautiful shop, Foxgloves Alley, while selecting fibers for the projects—and for her invaluable advice and encouragement throughout the entire project.

- to Candie Frankel for her painstaking attention to detail when checking (and rechecking) the manuscript, and for her superb organizational skills in pulling all of the pieces together and making them fit.

- to Judy Love for her flawless Stitch Library illustrations and to Lorraine Dey for drawing resplendently impeccable step-by-step illustrations and project diagrams.

- to Susan Raymond for her brilliance in photographic styling and to Bobbie Bush for capturing it on film.

- to Leslie Haimes for taking the text and photographs and so artfully designing such a magnificent book.

- to Manon Kavesky for her visionary creativity in marketing *The Embroidered Home*.

Finally, and most importantly, thank you to my family for putting up with all the disarray so I could complete this project. Your constant encouragement and reassurance, coupled with your unconditional love, mean more than you'll ever know.

About the Author

Ellen Moore Johnson is a designer, writer, and teacher who specializes in plain and fancy needlework, with an emphasis on surface embroidery. She began her adventure in the needle arts at the ripe old age of seven under the tutelage of her grandmothers. (Her first project was a dime store dresser scarf that she proudly embellished with harvest gold embroidery floss!)

A former needlework shop owner, Ellen brings more than fifteen years of teaching experience to her students. Her articles have appeared in numerous publications, and she is an advocate for the preservation and perpetuation of needlework as a fine art through her memberships in a variety of guilds and associations.

Ellen currently lives with her family in Tuscaloosa, Alabama. She considers herself incredibly blessed to have been able to pursue her lifelong hobby as a career.